Recent years have seen great interest in exploring universal gravitation both by astronomical and geophysical observations and by experiments in terrestrial laboratories. This book is about laboratory experiments, the reasons for doing them and the difficulties they present.

Gravitation is a very weak force and experiments will be useful only if they are at the limits of sensitivity and mechanical measurements. They must be imaginatively designed and carefully performed. They are especially afflicted by noise and other disturbances, the effects of which the authors analyse in detail, as they do issues in the design of experiments. Critical accounts are given of experiments for testing the inverse square law and the principle of equivalence and for measuring the constant of gravitation.

The book will be of value to graduate students, research workers and teachers engaged in any precise measurements or in theoretical or experimental studies of gravitation, who wish to understand the problems of laboratory experiments in this and similarly demanding fields.

GRAVITATIONAL EXPERIMENTS IN THE LABORATORY

GRAVITATIONAL EXPERIMENTS IN THE LABORATORY

Y. T. CHEN

Fellow of Corpus Christi College, University of Cambridge
Professor, Institute for Advanced Studies, University of Malaya

ALAN COOK

Master, Selwyn College, University of Cambridge

CAMBRIDGE
UNIVERSITY PRESS

CAMBRIDGE UNIVERSITY PRESS
Cambridge, New York, Melbourne, Madrid, Cape Town, Singapore, São Paulo

Cambridge University Press
The Edinburgh Building, Cambridge CB2 2RU, UK

Published in the United States of America by Cambridge University Press, New York

www.cambridge.org
Information on this title: www.cambridge.org/9780521391719

© Cambridge University Press 1993

First published 1993
This digitally printed first paperback version 2005

A catalogue record for this publication is available from the British Library

Library of Congress Cataloguing in Publication data

Chen, Y. T.
Gravitational experiments in the laboratory/Y. T. Chen, Alan Cook.
p.   cm.
Includes bibliographical references and index.
ISBN 0 521 39171 7
1. Gravity–Measurement.   2. Gravity–Experiments.   I. Cook, Alan H.   II. Title.
QC178.C645   1993
531'.14'078–dc20   92-7538   CIP

ISBN-13 978-0-521-39171-9 hardback
ISBN-10 0-521-39171-7 hardback

ISBN-13 978-0-521-67553-6 paperback
ISBN-10 0-521-67553-7 paperback

Contents

Contents

Preface

From the time that Newton first proposed that there was a universal force of gravity inversely proportional to the square of the distance between two point masses, there have been recurrent investigations of how far that rule was correct, and many different alternative forms have been suggested. The other assumption that Newton made, that the force of gravity did not depend on the chemical composition of bodies, has also been questioned from time to time; Newton himself carried out the first experimental test of what has become known as the weak principle of equivalence. It has often been suggested that some apparently anomalous behaviour in celestial mechanics should be ascribed to a failure of the inverse square law; indeed Clairaut developed the first analytical theory of the motion of the Moon because of discrepancies between Newton's theory and observation that might have been due to an inverse-cube component of the force. As with all subsequent studies before general relativity, careful analysis showed that the effects were consistent with the inverse square law. General relativity predicts a small deviation from the inverse square law close to very massive bodies, a deviation that has been confirmed by careful observation.

The motions of celestial bodies about each other are, with very minor exceptions, unable to reveal any departure from the weak principle of equivalence; if such departures are to be detected, they must be sought in laboratory experiments or geophysical observations. It has been argued that geophysical observations show failures of the simple Newtonian law of gravitation, but the most recent studies have not confirmed earlier results and are wholly consistent with the inverse square law and the weak principle of equivalence. Experiments in the laboratory, that is to say, at distances of no more than a few metres, have been performed to check the inverse square law at short distances and to examine the weak principle of equivalence, and again the outcome is that no deviations have been found from either. We may say that, at the present time, no departure from the predictions of general relativity has been confirmed and that general relativity remains the best theory of gravitation that we have.

None the less the nature of gravitation remains an enigma. Is it simply a matter of the geometry of space-time, or is it related to deeper physical symmetries as are the other principal forces of nature? The force is extremely weak and seems independent of any other physical circumstance, so that experimental investigations are very difficult, and up to now they have, as has just been said, confirmed the inverse square law and weak equivalence. Experiments on gravitation in the laboratory are therefore a great challenge to technique and imagination in design and performance, and many able experimenters have accepted that challenge over the last four decades, partly to examine the predictions of theories other than general relativity, partly to follow up indications from geophysical studies that there might have been failures of general relativity. As in geophysical studies, general relativity holds the field.

The considerable effort that has been spent on laboratory investigations should not be thought nugatory. The effects looked for are very small and push experimental design, technique and analysis to the limits. Consequently the general art of precise measurement has been greatly advanced by the search for anomalous effects, as well as by attempts to improve the precision of the measurement of the constant of gravitation, a constant that is still only poorly known. We have therefore thought it useful to describe the principal experiments that have been done in recent years, placing them in the historical context of earlier work and indicating possible future developments. We draw particular attention to the limitations set by noise and extraneous disturbances, for those are common to all very precise measurements and not just to experiments on gravitation. Experimental physicists in general, and not only those interested in gravitation, will, we trust, find value in our accounts of the very precise experiments that have been done and in the emphasis that we place on the importance of imaginative design carefully thought through. Theorists also, we hope, may find it useful to have a general critical account of the experimental evidence relevant to their investigations.

It seems that many programmes of laboratory investigations of gravitation are drawing to a close and that new ideas for experiments are needed if there is to be any prospect of detecting deviations from general relativity. The only realistic hope for radically different experiments may well lie in doing them in spacecraft, but while various experiments have been proposed, none has yet been accepted into a flight programme. Now therefore seems a suitable time to bring together the methods and results of recent work, giving, as they have, a generally negative answer to the question whether deviations from general relativity can be detected in laboratory experiments, but also having generated important advances in experimental technique.

There are many publications about the study of gravitation. We have been greatly helped by the publication of two bibliographies, one for the constant of gravitation (Gillies, 1987) and the other (Fischbach *et al.*, 1992) for work relevant to a possible 'fifth force'.

We have been much indebted in our studies of gravitation over the years to many colleagues in Cambridge and North America and mention, in particular, Dr P. L. Bender, Professor Sir Brian Pippard, Dr J. Faller, Dr G. Gillies, Dr A. J. F. Metherell, Dr Riley Newman, Dr C. Speake and Dr T. J. Quinn. One of us (A. H. C.) owes an especial debt to the late Professor Antonio Marussi, a friend and colleague over many years, whose untimely death was a great loss to geophysics and the study of gravitation. We take great pleasure in thanking those who have looked after us at Cambridge University Press, notably our editor, Dr S. Capelin, and our copy-editor, Sheila Champney.

1

Introduction

Three hundred years after Newton published *Philosophiae Naturalis Principia Mathematica* the subject of gravitation is as lively a subject for theoretical and experimental study as ever it has been (Hawking & Israel, 1987). Theorists endeavour to relate gravity to quantum mechanics and to develop theories that will unify the description of gravity with that of all other physical forces. Experimenters have looked for gravitational radiation, for anomalies in the motion of the Moon that would correspond to a failure of the gravitational weak principle of equivalence, for deviations from the inverse square law and for various other effects that would be inconsistent with general relativity. The cosmological implications of general relativity continue to be elaborated and various ways of using space vehicles to test notions of gravitation have been proposed. In particular, the last three decades have seen a considerable effort devoted to applying modern techniques of measurement and detection of small forces to experiments on gravitation that can be done within an ordinary physics laboratory, and it is those that are the subject of this book.

Our scope is indeed quite restricted. It is concerned with experiments where the conditions are under the experimenter's control, in contrast to observation, where they are not. It is concerned with experiments that can be done within a more or less ordinary-sized room, that is to say, the distances between attracting body and attracted body do not usually exceed a few metres and may often be much less, while the masses of gravitating bodies are of the order of kilograms or much less. Apparatus can be placed in carefully controlled enclosures and left undisturbed for long periods. Such conditions may be contrasted with those of geophysical experiments on the inverse square law of gravitation in which the value of gravity has been measured in boreholes or on high towers where the observer cannot control what he observes.

1

Why should anyone do laboratory experiments on gravitation, especially when we know from celestial mechanics that, on the large scale, gravitation is perfectly adequately described in terms of general relativity and almost always by the Newtonian approximation? To answer that question and to gain some idea of the precision for which we should aim in laboratory experiments, we must look briefly at the present state of theory, observation and experiment about gravitation.

According to Newton's law of gravitation the forces between two bodies of masses M_1, M_2 at a separation r, and of negligible extension, is $GM_1 M_2 r^{-2}$, and the experimenter might start by asking questions about the elements of that formula: is it really r^{-2}, does it matter what the composition of the masses is, what happens if the masses are not in vacuum but there is some other material between them, is G really a constant, or does it depend on time or where you are; and then at another level, does the force depend on velocity or on direction? According to general relativity, the force does vary very nearly as r^{-2} and nothing else has any effect. Celestial mechanics is entirely consistent with an inverse square law of force, apart from small deviations predicted by general relativity, but that cannot be taken to be a proof of general relativity, for the inverse square law might arise in other ways. Proofs of general relativity must be sought elsewhere. The motions of the perihelia of Mercury, Venus and the Earth (Duncombe, 1956) constitute one of the classical tests, the others being the deflection of light by the Sun and the gravitational red shift. The deflection of radio waves by the Sun can now be established more precisely by long base line radio interferometry (Fomalmont & Sramek, 1977) than can the optical deflection, while the time delays of radio signals passing near the Sun have also been measured with very high precision (Shapiro *et al.*, 1972a; Anderson *et al.*, 1975; Reasenberg *et al.*, 1979). In addition, radar observations to Mercury have greatly improved the determination of the motion of the perihelion (Shapiro *et al.*, 1972a; Shapiro, 1980). The observational basis of general relativity has been greatly strengthened in recent years, to a large extent as the result of the Apollo and planetary space programmes of the United States of America.

General relativity gives a geometrical account of gravitation: small, compact test bodies move along geodesics of space-time so that their accelerations are determined solely by that geometry and do not depend at all upon the nature of the material. Other bodies have the sole effect that they alter the space-time geometry and do not introduce any shielding as may occur with electromagnetic forces. In the neighbourhood of an isolated compact massive body the metric of space-time is the Schwarz-schild metric,

$$ds^2 = (1 - 2GM/rc^2)(dx^0)^2 - (1 - 2GM/rc^2)^{-1} dr^2 - r^2 d\theta^2 - r^2 \sin^2\theta\, d\phi^2,$$

(1.1)

where ds is the infinitesimal space-time interval, x^0 is ct and θ and ϕ are the colatitude and azimuthal angle.

When GM/rc^2 is much less than 1, the metric may be taken to be

$$\mathrm{diag}\,[(1 - 2GM/rc^2),\ -1,\ -1,\ -1],$$

(1.2)

while the equation of motion reduces to

$$\frac{d^2 x}{dt^2} = \frac{d}{dr}\left(-\frac{GM}{r}\right),$$

(1.3)

which is just Newton's law with a potential of the form $-GM/r$.

When the finite magnitude of GM/rc^2 has to be taken into account, the factor $(1 - 2GM/rc^2)^{-1}$ by which dr^2 is multiplied has to be included and that leads to an additional term, proportional to r^{-2}, in the potential and thus to the motions of the perihelia of Mercury and other planets close to the Sun. So far as is known, all celestial bodies move in orbits consistent with general relativity and the very great majority in orbits according to the inverse square law. It is by no means a simple matter to establish that conclusion. If there were just two bodies attracting each other according to the inverse square law, then they would move in confocal elliptic orbits, but no such isolated systems exist and the orbits of all celestial objects are perturbed to a greater or lesser degree by the attractions of other bodies. The most obvious case is that of the orbits of the Moon and Earth about each other, which are strongly affected by the attraction of the Sun (Cook, 1988 b); the planets also show substantial mutual perturbations. None the less the inverse square law, modified where necessary by the additional term of the Schwarzschild metric, has so far satisfactorily accounted for all such perturbed motions. Closer to the Earth, it is found that the value of GM for the Earth derived from the motions of artificial satellites is within 1 part in 10^7 of that derived from measurements of gravity over the surface of the Earth, when account is taken of the deviation of the Earth's shape from a sphere, the inhomogeneous distribution of the Earth's density and the mass of the atmosphere. Further, no effects of preferred directions in space have been detected. From the time of Clairaut onwards, a continuing feature of celestial mechanics has been attempts to account for some perturbation by a force additional to or inconsistent with the inverse square law, but all such have proved illusory and attributable to mistakes in the dynamical theory. Clairaut's initial lunar theory was the first (Cook, 1988 b).

While celestial mechanics gives good grounds for the validity of the inverse square law at celestial distances, it can tell us nothing about the influence, if any, of the material of celestial objects. Most are in any case mainly hydrogen, but we have no independent estimate of the masses apart from the parameters of the orbits and so dependence on material would only make a slight change in what the mass of a body was thought to be. There is one effect that has been suggested – a perturbation in the lunar orbit arising from different accelerations of the Earth and the Moon under the attraction of the Sun (Nordtvedt, 1968) but the expected effect would be very small. It has been sought in the results of laser ranging to the retroreflectors placed on the Moon in the Apollo programme (Shapiro *et al.*, 1976; Williams *et al.*, 1976) but no effect has been found and the uncertainties of the results place a limit of 7 parts in 10^{12} on deviations from the weak principle of equivalence as between the Sun and the Moon. It is a limit very similar to that set by laboratory experiments (Chapter 4).

Deviations from the inverse square law could arise in two ways: a true departure of gravitation itself, or additional non-gravitational forces. If it were the first, the deviation would be the same for all materials, but if there were additional forces that were confused with gravitation, then the effects might depend on the nature of the material. An obvious cause of the first situation would be that the gravitational force was carried by a particle that had mass instead of being carried by a massless particle such as the photon. In that case the gravitational potential would have the form

$$-(GM/r)(1+\alpha \, \mathrm{e}^{-\mu r}), \qquad (1.4)$$

and one of the aims of experiment in the laboratory is to set limits on such deviations from the inverse square law.

Deviations from the inverse square law have also been sought on the geophysical scale. The constant μ is the inverse of a scale length. If r were very great compared with the scale length, the law of force would be effectively inverse square; that is the situation in celestial mechanics. On the other hand, if r is small compared with the scale length, the constant in the law would not be G but $(1+\alpha)G$; that seems to be the situation in laboratory experiments, where again no deviation from the inverse square law has been established.

At distances comparable with $1/\mu$, there would be a change of the apparent value of the constant of gravitation from G to $(1+\alpha)G$, and that has been sought in geophysical experiments. There were initially some suggestions of real effects in measurements of gravity in boreholes but it seems now that the proper allowances had not been made for the attraction of the irregular ground above the boreholes. Again, there had been indications that the gradient of gravity as measured above ground on a tall tower was not consistent with the inverse square law, but further measurements, in which careful account was taken of such matters as the vertical acceleration due to the horizontal sway of the tower and the topography of the ground at the surface, show no deviation from the predictions of the inverse square law (Cruz *et al.*, 1991; Fischbach & Talmadge, 1992).

The rather poor precision of about 1 part in 10^4 with which the inverse square law for gravity can be established in the laboratory is in marked contrast to that for electrostatics, about 1 part in 10^{16} (Williams, Faller & Hill, 1971). One reason is the very great sensitivity of electrical measurements as compared with mechanical measurements, the other is the fact that electrical detectors can be completely enclosed within a conducting Faraday cage, whereas it is not possible to build a completely enclosed gravitational Faraday cage and still have access to a mechanical detector. Such contrasts between the precision and possibilities of electrical and mechanical measurements are at the heart of the experimental problems of gravitational experiments.

The conclusion drawn from general relativity that a gravitational acceleration is independent of the material of the moving object, is known as the weak principle of equivalence. It is one of the most securely established facts of physics (see Haugan & Will, 1976; Haugan, 1979), but in recent years its validity has been questioned, initially because of the suggestion that kaons might produce a force proportional to the baryon number instead of the total mass of a body in which the binding energy is included in the total mass. Fischbach *et al.* (1986) therefore re-examined the results of Eötvös, Pekar & Fekete (1922, 1935) and of Renner (1935), who had compared the attractions of the Earth on bodies of various materials and had concluded that the weak principle of equivalence was valid for those materials to within 1 part in 10^9. Fischbach and his colleagues claimed that there was a residual deviation from the principle but their analysis has been criticized by others (see for example, Keyser, Niebauer & Faller, 1986; Kim, Klepacki & Hinze, 1986; Cook, 1988a), and subsequent experiments have shown no effect.

Various other deviations from the simple Newtonian law have been sought, for the most part without success (see Will, 1987; Cook, 1987a,

1988a). If there were a preferred direction in the universe, it might show up as a variation of gravity at a point on the Earth as the Earth rotated about its axis. There are indeed such variations but so far they can be perfectly well explained by the known tidal forces acting on the Earth and the deformations that the Earth undergoes as a result (Will, 1981). Similarly, no preferred direction for inertia has been found (Hughes, Robinson & Beltran-Lopez, 1960; Drever, 1961; Phillips, 1965a, 1987).

The weak principle of equivalence is concerned with the matter of an attracted test object. Experiments have also been done to see if the gravitational force exerted by a massive body depends on its composition. Thus Kreuzer (1968) placed a cylinder of polytetrafluoroethylene (Teflon) composed of 75 per cent of fluorine in a tank containing a mixture of trichloroethylene and dibromomethane (74 per cent of bromine) adjusted in composition to have the same density as Teflon. The Teflon cylinder was moved from side to side in the tank and a torsion balance was used to detect any change of attraction on a test mass. None was found, limiting any difference between the attractions of Teflon and the liquid to 1 part in 10^5.

Gravitational shielding has attracted a certain amount of attention. Can gravitational forces be shielded like electrical or magnetic forces? Thus Austin & Thwing (1897) using a torsion balance similar to that of Boys (1895) found that screens of lead, zinc, water, mercury, alcohol and glycerine, when placed between attracting bodies, had no effect to 1 part in 500 (see Braginsky, Radenko & Rukman, 1962). Majorana (1957) has on the other hand, claimed on many occasions to have found a definite effect in experiments in which a sphere of lead was surrounded by screens of 114 kg of mercury or 9880 kg of lead. Any shielding effect would mean that gravitational forces are not simply additive and Saxl & Allen (1971) have indeed claimed that the forces exerted by the Sun and the Moon upon a torsional pendulum were not additive. Harrison (1963), however, discussed observations of the tidal variation of gravity and found that the variation of gravity when the Sun was below the horizon of the site of an observation was in no way anomalous. A stringent limit was also placed on any failure of additivity by Slichter, Caputo & Hager (1965) through analysis of Earth tide observations during the solar eclipse of 1961. The same eclipse was observed by Caputo (1962) at Trieste, where totality occurred when the Sun and the Moon were almost due west of Trieste and at an altitude of about 13°. Two horizontal pendulums of very long period had been set up near Trieste in the Grotta Gigante, one responding to a north–south variation of gravity, the other to an east–west variation.

Neither showed any anomaly in the tidal variations of gravity as the Sun passed through totality behind the Moon. The eclipse experiments hardly prove that there is no shielding but they do show that the attractions of the Sun and the Moon are purely additive.

There have often been speculations that the constant of gravitation may change with time (Dyson, 1972). Dirac (1937) in particular advanced the so-called large number hypothesis according to which certain non-dimensional combinations of fundamental constants would be proportional to the age of the universe. In that case, G would have changed by an order of magnitude since the beginning of the universe. That is far too small a rate to be detected by laboratory measurements of the constant, but it might be found from fundamental astronomical observations (see Shapiro *et al.*, 1972b; van Flandern, 1981). What would a change of G imply and how might it be found? If the question is to be meaningful, it must be possible to determine G in terms of other physical standards, such as the speed of light, the atomic standard of frequency, and combinations of electrical quantities such as h/e, but excluding the mechanical standard of mass. Although that cannot be done to realistic accuracy in the laboratory, it is possible in principle to use astronomical measurements of time. Our fundamental standard of time and frequency is nowadays an atomic standard and the frequency of the electrical signal corresponding to transitions in the atomic system depends on atomic constants such as h and e. Gravitation does not come into the structure of the atom nor the frequencies of transition. Astronomical time scales, such as ephemeris time, are based on some well-defined orbital period – ephemeris time is based on the orbit of the Earth about the Sun. Now the angular velocity ω of a planet about its primary is given by

$$a^3\omega^2 = GM, \qquad (1.5)$$

where a is semi-major axis of the orbit, and if the angular momentum is constant, equal to L, then

$$a^2\omega = L, \qquad (1.6)$$

hence

$$\omega = (GM)^2/L^3. \qquad (1.7)$$

If then it can be assumed that L and M are not changing, a comparison of ephemeris and atomic time should show a change in G. The best limits so far have been set by spacecraft measurements in which distances as well as times can be measured (see Will, 1987).

Nothing is said in this book about the detection of gravitational radiation. It is a rather special topic and in many ways the experimental problems are distinct from those discussed here (Drever, 1983). A real detection of gravitational waves would of course give strong support to general relativity, as well as opening up a whole new field of astronomical investigation. So far no signals have certainly been detected on Earth, but the binary pulsar discovered by Hulse & Taylor (1975) has provided strong evidence that its orbital period of about 8 h is decreasing at the rate of 8 parts in 10^{17} per second on account of loss of energy by gravitational radiation (Haugan, 1985). It has thus confirmed the occurrence of gravitational radiation and also general relativistic celestial mechanics.

The current state of our knowledge of gravitation is that nothing so far has been found inconsistent with general relativity. It may be in fact that gravity really is nothing but a geometrical property of the universe, but if it is not, if the aim of unifying it with other forces of nature is a realistic one, then that must be done in such a way that maintains the purely geometrical character to a very high order (see Richardson, 1910, 1922). It has been seen in this brief introduction that modern astronomical techniques such as radar observations of the planets and observations from and to space vehicles have helped to bring us to that understanding, but very careful experiments in the laboratory have also been important (Braginsky, Caves & Thorne, 1977), and it is the purpose of this book to give an account of the most important problems, methods and results of those experiments.

2

The linear oscillator driven by thermal noise and with electrical damping

2.1 Introduction

It is well known that because gravitational forces are very weak, the experimental study of gravitation is difficult and despite recent advances in techniques of data handling and analysis, it is still subject to great limitations.

The general method of measuring a gravitational force is to observe the amplitude or frequency of a linear harmonic oscillator which can be subject to the external force. The oscillator could be a torsion balance, a beam balance, a simple pendulum or a resonance antenna (see Cook, 1988a). The limit set to the measurement by noise and other disturbances is the subject of this chapter and Chapter 3.

A linear harmonic oscillator is an oscillator described by the equation

$$m\frac{d^2x}{dt^2} + H\frac{dx}{dt} + \lambda x = 0, \qquad (2.1)$$

where m is the mass, H is the damping factor, λ is the rigidity. For measuring a very weak force, the oscillator must have a relatively high Q (the mechanical quality).

The so-called linear harmonic oscillator is neither strictly linear nor harmonic. It has various imperfections which will disturb the amplitude and frequency and, in turn, will limit the smallest force it can detect (Chen & Cook, 1990b). That force we call the least detectable force.

Since, as stated in Chapter 1, this book is mainly concerned with practical experiments, a general discussion of all possible sources of disturbance will not always be necessary. We give a general account when that is possible and helpful, but when a restricted treatment is adequate, we will usually discuss the torsion balance because that is the detector most often used in experiments on gravitation.

Two kinds of disturbance affect the response of the oscillator. The first comprises those external sources which can, to a large degree, be removed: for example, electromagnetic forces, temperature drift, extraneous gravitational attraction or seismic vibration of the ground. Their effects can be greatly reduced either because the oscillator can be isolated from the sources of the disturbance, or because the effect can be removed by regression on independent measurements of the specific disturbance. The second kind of disturbance is intrinsic noise, which cannot be removed from the motion of the oscillator and which sets the fundamental limit to the measurement. Examples of this kind of noise are: the thermal fluctuation of the oscillator at a certain temperature and the shot noise in the optical detection system for a certain level of power. In the measurement of gravitational force, thermal noise sets a fundamental limit. Although the effect may be reduced by working at low temperature, the reduction is only proportional to the square root of the absolute temperature, while cooling the system may give rise to serious problems of mechanical vibration.

Sources of disturbance may also be classified by their spectra and internal correlation. Thus thermal noise has a white spectrum and there is no internal correlation, whereas seismic noise has a peaked spectrum and there is autocorrelation. In the following section we consider the response of a linear oscillator to an uncorrelated random noise source. It should be emphasized at the start that although the noise source has a white spectrum and is uncorrelated, the response of the linear oscillator will be peaked and will show autocorrelation.

2.2 Response of linear oscillator to thermal noise

If a force $f(t)$ acts on a linear harmonic oscillator the equation of motion followed by the test mass m of the oscillator, is

$$m\frac{\mathrm{d}^2x}{\mathrm{d}t^2} + H\frac{\mathrm{d}x}{\mathrm{d}t} + \lambda x = f(t). \tag{2.2}$$

We assume that $f(t)$ represents the sum of an applied gravitational force and a thermal noise driving force and we consider how the displacement or the period of the linear oscillator changes in the presence of a thermal noise force (see Mazur, 1940; Milatz & Vanzolingen, 1953; Braginsky, 1968).

Because the oscillator is linear, its responses to the applied force and to the thermal driving force are additive. At first sight it might seem that the response to thermal noise could be obtained from the equipartition formulae of Einstein:

$$\tfrac{1}{2}\lambda \,\overline{x_n^2} = \tfrac{1}{2}k_b\, T,$$ (2.3)

$$\tfrac{1}{2}m\,\overline{\dot{x}_n^2} = \tfrac{1}{2}k_b\, T.$$ (2.4)

However, those formulae are only valid if the system has attained a statistical equilibrium state and that is usually not the case in the practical use of linear oscillators for gravitational measurements. To obtain high sensitivity the Q is made very large which means that the time for transient disturbances to die away is usually much longer than the duration of a practical experimental measurement. Consequently, the oscillator will be in a non-equilibrium state and the transient forms of the solution of the equation of motion are required. Accordingly, in this chapter we derive the transient solutions of the equation of motion of a linear oscillator driven by random forces as well as by a gravitational force of arbitrary period.

The transient noise response discussed is not the same as the Brownian motion of a free particle, which is subject to viscous dissipation but not to a restoring force. Although the driving force as described below is totally random and covers a frequency spectrum ranging from zero to infinity, the response of the oscillator has a limited range of frequency because the oscillator is harmonically bound and has a high mechanical quality factor. Consequently, over a short time scale the displacement of the noise response may be approximately sinusoidal, but after a long time all possible combinations of phase and frequency may occur (Pippard, 1978).

Put $H/m = \beta$, $\lambda/m = \omega_0^2$ and $f(t)/m = F(t)$, so that eqn (2.2) becomes

$$\frac{\mathrm{d}^2 x}{\mathrm{d}t^2} + \beta\frac{\mathrm{d}x}{\mathrm{d}t} + \omega_0^2 x = F(t).$$ (2.5)

The acceleration $F(t)$ is taken to be the sum of two parts, $F_n(t)$ of the noise and $F_s(t)$ of the signal. Because the system is linear, we may write the response $x(t)$ as

$$x(t) = x_n(t) + x_s(t) + x_i(t),$$ (2.6)

where $x_i(t)$ is the complementary function dependent on the initial conditions of the oscillator, $x_s(t)$ is the particular solution for the inhomogeneous equation with only $F_s(t)$ present and $x_n(t)$ is the particular solution for the inhomogeneous equation with only $F_n(t)$ present.

We ask first, what is the least detectable acceleration in the face of the random noise when the applied force is measured by the deflection of a linear oscillator? Generally speaking, the least detectable acceleration will depend upon three factors: the method of measurement, the time of observation and the difference of frequencies between the linear oscillator, which acts as a detector, and the applied force.

The solution of the equation of motion is expressed as a response function $x(t)$ giving the value of the displacement at a some arbitrary instant t. In practice, observation of the displacement may be made in two different ways: discrete and continuous recording. In some experiments, the effect of an applied force can be detected by making spot measurements of the response $x(t)$ at time t and comparing them with measurements at $t = 0$ when the force starts to act (of course, $x(t)$ may be measured several times at different instants to reduce the error in the mean). Such measurements, made point by point, we call discrete measurements. In other experiments, the displacement may be recorded continuously for a certain period of time or at very short time intervals and the data may then be integrated in an appropriate way to give a quantity that should be proportional to $F_s(t)$ – that is, one forms a functional of the function $x(t)$. For example, if $F_s(t)$ is supposed to be constant in time, the appropriate functional would be the integral of $x(t)$ over a certain integral number of free periods of the oscillator and the functional would then be proportional to $F_s(t)$. Such measurements will be called continuous measurements. The least accelerations detectable by discrete and continuous measurements may be different; both will be discussed.

The frequency p of the signal is another important factor in the estimation of the least detectable acceleration, because the difference between the frequencies of the signal and the oscillator will greatly affect the expected value. The response of the oscillator to a signal with arbitrary period will be derived but, in practice, two cases are particularly important: a period near resonance ($p \approx \omega_0$) and a constant force ($p = 0$).

If a signal is periodic, it is best detected with a resonant oscillator which gives the minimum detectable signal. Many gravitational experiments may be carried out by the resonant method even if the frequency of the oscillator differs slightly from the frequency of the external gravitational force and the resonant condition provides a good standard against which to estimate the effect of noise in the experiment.

The case of $p = 0$ corresponds to a constant applied force. That is also very common in gravitational experiments, particularly the determination of the gravitational constant and the verification of the inverse square law, or the test of the presence of other forces. At the present stage of technical development, the precision of such measurements is not limited by thermal noise, but nevertheless a theoretical investigation of the ultimate limit set by thermal noise is valuable both for establishing the likelihood of the detection of certain effects and to indicate how current methods might be improved.

The least detectable signal will depend upon the observation time. It is usually supposed as a matter of principle that the longer the observation time, the smaller the signal that can be detected. As we shall show, that is not always so. We place no restriction on the time of observation, but two extreme time scales are particularly discussed:

$$t \ll 2/\beta \quad \text{and} \quad t \gg 2/\beta. \tag{2.7}$$

The first represents the most common situation and corresponds to an undamped oscillator with a very high Q and a very long relaxation time. The second corresponds to a situation where the system has reached an equilibrium state, again very common. It should be pointed out here that any damping force, whether mechanical, hydrodynamical, electrical or otherwise, necessarily introduces thermal fluctuations. If damping is by the viscosity of air at low pressure, high sensitivity necessarily entails a high Q and very long relaxation time. Electrical damping with a servocontrol system affords wider possibilities for rapid damping, but, as discussed below, will not introduce extra fluctuation forces. Therefore, the results obtained here for the response to noise when $t \gg 2/\beta$ can also be used for an electrically damped oscillator.

To obtain the solution of eqn (2.5), take $F_n(t)$ to be an arbitrary force and let the initial conditions be:

$$x_{t=0} = x_0, \tag{2.8}$$

$$\left(\frac{\mathrm{d}x}{\mathrm{d}t}\right)_{t=0} = u_0. \tag{2.9}$$

We suppose $F_n(t)$ to be the random acceleration corresponding to bombardment by the molecules of the surrounding gas. It is assumed to have the following properties:

(1) it has zero expectation at any given time t, that is,

$$\langle F_n(t) \rangle = 0, \tag{2.10}$$

where the mean value denoted by $\langle \ \rangle$ is an ensemble mean;
(2) the variance is a constant:

$$\langle F_n(t)^2 \rangle = D, \tag{2.11}$$

where D is a constant;

(3) the forces at two different times t_1 and t_2 are assumed to be uncorrelated,

$$\langle F_n(t_1) F_n(t_2) \rangle = D \delta(t_1 - t_2), \tag{2.12}$$

where δ is the δ-function and the mean value is again an ensemble mean.

The initial position x_0 and the initial velocity u_0, are the result of noise prior to $t = 0$ and so are unknown and have zero means. Hence, when the displacement at time t is evaluated, an ensemble average over u_0 has to be taken.

The solutions $x_i(t)$ and $x_n(t)$ of eqn (2.6) are

$$x_i(t) = e^{-\beta t/2} \left[x_0 \left(\cos \omega_1 t + \frac{\beta}{2\omega_1} \sin \omega_1 t \right) + \frac{u_0}{\omega_1} \sin \omega_1 t \right], \tag{2.13}$$

$$x_n(t) = \frac{1}{\omega_1} \int_0^t F_n(\xi) e^{-\beta(t-\xi)/2} \sin \omega_1 (t-\xi) \, d\xi, \tag{2.14}$$

where ω_1 is given by

$$\omega_1^2 = \omega_0^2 - \beta^2/4. \tag{2.15}$$

If the signal $F_s(t)$ is periodic, with an arbitrary frequency p, that is,

$$F_s(t) = C e^{ipt}, \tag{2.16}$$

then the Laplace transform of the equation of motion is

$$(s^2 + \beta s + \omega_0^2) \eta(s) = \phi(s), \tag{2.17}$$

where $\eta(s)$ is the Laplace transform of $x_s(t)$ and $\phi(s)$ is the Laplace transform of $F_s(t)$, namely

$$\phi(s) = \int_0^\infty C e^{ipt} e^{-st} \, dt = \frac{C}{s - ip}. \tag{2.18}$$

Let

$$s^2 + \beta s + \omega_0^2 = (s - \sigma_1)(s - \sigma_2) \tag{2.19}$$

with

$$\sigma_1, \sigma_2 = -\tfrac{1}{2}\beta \pm i\omega_1 \tag{2.20}$$

and

$$\sigma_1 \sigma_2 = \omega_0^2. \tag{2.21}$$

Then

$$\eta_s(s) = \frac{C}{(s - ip)(s - \sigma_1)(s - \sigma_2)} \tag{2.22}$$

and the real response $x_s(t)$ is

$$x_s(t) = \frac{C[(\omega_0^2 - p^2)\cos pt - p\beta\sin pt]}{(\omega_0^2 - p^2)^2 + p^2\beta^2}$$

$$- Ce^{-\beta t/2}\frac{[(\omega_0^2 - p^2)\cos\omega_1 t - (\beta/2\omega_1)(\omega_0^2 + p^2)\sin\omega_1 t]}{(\omega_0^2 - p^2) + p^2\beta^2} \qquad (2.23)$$

(also see Pippard, 1978). If F_s is constant, p is zero, and

$$x_s(t) = \frac{C}{\omega_0^2}\left[1 - e^{-\beta t/2}\left(\cos\omega_1 t - \frac{\beta}{2\omega_1}\sin\omega_1 t\right)\right]. \qquad (2.24)$$

If p is equal to ω_1 and also if β is very small compared to ω_0, then

$$x_s(t) = -\frac{C}{\omega_1\beta}(1 - e^{-\beta t/2})\sin\omega_1 t. \qquad (2.25)$$

If the applied force is purely harmonic so that p is real and there is no decay, then ip can never be equal to the complex speeds σ_1 and σ_2 and the poles of the integrand in the inversion of the Laplace transforms are always distinct. Consequently $x_s(t)$ has no component proportional to time.

The calculation of the fluctuation of the linear oscillator driven by noise is a problem that has been studied by many authors, e.g. Uhlenbeck & Ornstein (1930), Kramers (1940), Wang & Uhlenbeck (1945) and Prigogine (1962). In gravitational experiments we usually employ a lightly damped oscillator and record its displacement or period before it has reached equilibrium. We wish to estimate how the deflection is affected by the Brownian motion of the oscillator, or what limit is set by Brownian motion to the measured period of the oscillator in circumstances where equilibrium values for the noise displacement and velocities are not applicable, and for those reasons our approach differs from that of previous authors.

2.3 Criteria for least detectable force in amplitude measurement

Any force applied to a linear oscillator may change the amplitude or the equilibrium position of the oscillator and it may change the period of oscillation. We begin by establishing criteria for the least force that can be detected by a change of amplitude or deflection. Those criteria should be as close as possible to real experimental situations.

The force may be estimated from the response $x(t)$ of the linear oscillator at a particular time t after the force is applied. Alternatively, it may be estimated from some average value of the response, that is from the integral over a span of time, namely the functional

$$\overline{x(t)} = \frac{1}{t_2 - t_1} \int_{t_1}^{t_2} x(t)\,G(t)\,dt, \qquad (2.26)$$

where $G(t)$ is an appropriate weighting function.

The form chosen for $G(t)$ will depend on the form of $x(t)$ as a function of time, whether a constant or a periodic function, or something else. The form of the function $G(t)$ and the instants t_2, t_1 will be determined in such a way that the functional should equal the expected value of the measurement. For example, when the span of $t_2 - t_1$ equals a few periods of free oscillation of the pendulum, two obvious choices for the function $G(t)$ are:

(1) when the applied force is periodic with frequency ω

$$G(t) = \begin{cases} +1 & \text{when} \quad \dfrac{2n\pi}{\omega} \leqslant t < \dfrac{2(n+\frac{1}{2})\pi}{\omega}, \\[2ex] -1 & \text{when} \quad \dfrac{2n\pi}{\omega} \leqslant t < \dfrac{2(n+1)\pi}{\omega}, \end{cases} \qquad (2.27)$$

n is an integer;

(2) when the applied force is constant,

$$G(t) = 1. \qquad (2.28)$$

To establish the criteria for the least force detectable in the experiment, we estimate significant differences between $x(t)$ for discrete and continuous observations and some reference value, x_0.

In the case of a discrete observation, consider first the application of a steady force from the time $t = 0$ when the deflection of the pendulum is x_0. How small a force can be detected by observing the instantaneous deflection at a later time t? We adopt the criterion used by Braginsky & Manukin (1977) namely that for a signal to be detectable in any observation at time t

$$(x_s - x_0)^2 \geqslant \langle (x_i + x_n - x_0)^2 \rangle. \qquad (2.29)$$

The reference value x_0 enters the criterion because the deflection must be measured from some origin, and an obvious one to choose is the initial deflection at the time when the force was first applied. The particular value of x_0 is of course arbitrary but may always be taken to be the origin from which deflections are measured. However, because both x_0 and the initial velocity u_0 are determined by noise forces on the pendulum at times prior to $t = 0$, we cannot assume that x_0 and u_0 are both arbitrary, so that u_0

cannot be assumed to be zero even though x_0 can be taken to be the arbitrary zero of position.

We should also consider the initial phase of the signal, so that instead of eqn (2.16), $F_s(t)$ should be written as

$$C \, e^{ipt+\phi_0}, \tag{2.30}$$

where ϕ_0 is the initial phase difference between the applied force and the motion of the oscillator. When $t = 0$, ϕ_0 should correspond to the initial value of x_0, otherwise the comparison would be very complicated, but it is always possible to take both

$$x_0 = 0, \tag{2.31}$$

and

$$\phi_0 = 0. \tag{2.32}$$

Therefore, the criterion for the detection of a force by the displacement at a time t after applying the force is

$$x_s^2 \geqslant \langle (x_n + x_i)^2 \rangle. \tag{2.33}$$

The ensemble average of the product $x_n x_i$ vanishes because of condition (2.12) and therefore the criterion becomes

$$x_s^2 \geqslant \langle x_n^2 \rangle + \langle x_i^2 \rangle. \tag{2.34}$$

If a periodic force is applied, the same criterion can be used except that t should be chosen to correspond to the moments when the amplitude of the oscillator attains maximum and minimum values.

Now suppose that instead of making an instantaneous observation of the deflection, we form a weighted average over some interval. If we apply a force that is constant in time, the appropriate average is (by eqn (2.28))

$$\overline{x(t)} = \frac{1}{\Delta t} \int_{t_1}^{t_2} x(t) \, dt, \tag{2.35}$$

where

$$\Delta t = t_2 - t_1, \tag{2.36}$$

the times t_2 and t_1, being chosen so that the value of $\overline{x(t)}$ is simply proportional to C, the amplitude of the applied force (see eqn (2.16)).

But

$$\frac{1}{\Delta t} \int_{t_1}^{t_2} x(t) \, dt = \frac{1}{\Delta t} \int_{t_1}^{t_2} x_s \, dt + \frac{1}{\Delta t} \int_{t_1}^{t_2} x_n \, dt + \frac{1}{\Delta t} \int_{t_1}^{t_2} x_i \, dt, \tag{2.37}$$

and so the criterion for the detection of a force will be

$$\left[\frac{1}{\Delta t} \int_{t_1}^{t_2} x(t) \, dt \right]^2 \geqslant \left\langle \left[\frac{1}{\Delta t} \int_{t_1}^{t_2} x_n \, dt + \frac{1}{\Delta t} \int_{t_1}^{t_2} x_i \, dt \right]^2 \right\rangle, \tag{2.38}$$

which again by the condition of eqn (2.12) becomes

$$\left[\frac{1}{\Delta t}\int_{t_1}^{t_2} x(t)\,dt\right]^2 \geqslant \left\langle\left[\frac{1}{\Delta t}\int_{t_1}^{t_2} x_n\,dt\right]^2\right\rangle + \left\langle\left[\frac{1}{\Delta t}\int_{t_1}^{t_2} x_i\,dt\right]^2\right\rangle. \qquad (2.39)$$

If a periodic signal is to be found from a weighted average of the deflection, the weighting function $G(t)$ in the case of eqn (2.27) must be used and the criterion to be applied for the deflection of a periodic signal will be similar to that of eqn (2.39).

It should be mentioned that other functionals might be chosen as estimators, for example the difference of two weighted averages at different epochs

$$d = \frac{1}{\Delta t}\left[\int_{t_2}^{t_2+\Delta t} x(t)\,G(t)\,dt - \int_{t_1}^{t_1+\Delta t} x(t)\,G(t)\,dt\right]. \qquad (2.40)$$

It will be obvious how the methods of this section may be applied to more general estimators.

In calculating the ensemble average of the square of the response of some estimator of noise, it is necessary to allow for the fact that whereas noise at different times is uncorrelated, responses to noise at different times, when the measurement time is less than the relaxation time, are not uncorrelated. An inspection of eqn (2.14) shows that the integrals for x_n at different times t_1 and t_2, say, where t_2 is greater than t_1, contain the same noise function over the span of time from $t = 0$ to $t = t_1$; thus the responses $x_n(t_1)$ and $x_n(t_2)$ are correlated. In consequence, when ensemble means of estimators are calculated, the integrals over the noise must be evaluated explicitly; we may expect that continuous observations should depress the effect of noise considerably for a harmonically bound system which is in the non-equilibrium state.

2.4 Least detectable signal – discrete measurements

Consider the ensemble variance of the noise response according to the criterion given by eqn (2.34). From eqn (2.13) we have

$$\langle x_i^2\rangle = \frac{\langle u_0^2\rangle}{\omega_1^2}e^{-\beta t/2}\sin^2\omega_1 t. \qquad (2.41)$$

Then, from eqn (2.14), we have

$$\langle x_n^2\rangle = \frac{1}{\omega_1^2}\int_0^t\int_0^t \overline{F_n(\xi)\,F_n(\zeta)}\,e^{-\beta(t-\xi)/2}\,e^{-\beta(t-\zeta)/2}\sin\omega_1(t-\xi)\sin\omega_1(t-\zeta)\,d\xi\,d\zeta.$$

$$(2.42)$$

On inserting eqns (2.10)–(2.12) into the above equation, the variance of x_n is given by (Uhlenbeck & Ornstein, 1930)

$$\langle x_n^2 \rangle = \frac{D}{2\omega_1^2 \beta}(1-e^{-\beta t}) - \frac{D}{8\omega_1^2 \omega_0^2}(\beta - e^{-\beta t}\beta \cos 2\omega_1 t + 2\omega_1 e^{-\beta t}\sin 2\omega_1 t).$$

(2.43)

In the two equations (2.41) and (2.43), there are two constants to be determined, $\langle u_0^2 \rangle$ and D. The expectation of $\langle u_0^2 \rangle$ from eqn (2.4) is given by

$$\langle u_0^2 \rangle = \frac{k_b T}{m}.$$

(2.44)

The equipartition variance of $\langle x_n^2 \rangle$ is $k_b T/m\omega_0^2$, and combining that with the value given by eqn (2.43) at infinite time, we obtain

$$D = \frac{2k_b T\beta}{m}.$$

(2.45)

Inserting eqns (2.44) and (2.45) into (2.41) and (2.43) respectively, we find

$$\langle x_n^2 \rangle + \langle x_i^2 \rangle = \frac{k_b T}{m\omega_0^2}\left[1 - \left(\cos \omega_1 t + \frac{\beta}{2\omega_1}\sin \omega_1 t\right)^2 e^{-\beta t}\right].$$

(2.46)

The factor $(k_b T/m)^{\frac{1}{2}}$ enters all expressions for the least signal, and it is convenient to denote it by u_n; it is the rms velocity of a body of mass m in thermal equilibrium at temperature T; u_n for a typical mass of 50 g of a torsion pendulum is 10^{-10} m s^{-1} at room temperature.

Then

$$\langle x_n^2 \rangle + \langle x_i^2 \rangle = \left(\frac{\tau_0 u_n}{2\pi}\right)^2\left[1 - \left(\cos \omega_1 t + \frac{\beta}{2\omega_1}\sin \omega_1 t\right)^2 e^{-\beta t}\right],$$

(2.47)

where τ_0 is the free period of the linear oscillator.

With the expression for noise from eqn (2.47), the least detectable force can be calculated from the criteria obtained in section 2.3. Some cases commonly encountered in experiments are now considered.

(1) *Resonance.* The free period of the linear oscillator is the same as, or close to, the period of the applied force, and $1/\omega_1 < t \ll 2/\beta$. Under these conditions, from eqn (2.25),

$$x_s \approx \frac{Ct}{2\omega_1}\sin \omega_1 t.$$

(2.48)

We compare that value with the mean square fluctuation from eqn (2.47) which is

$$\langle x_n^2 \rangle + \langle x_i^2 \rangle = \left(\frac{\tau_0 u_n}{2\pi}\right)^2 \left[1 - \left(\cos \omega_1 t + \frac{\beta}{2\omega_1} \sin \omega_1 t\right)^2 (1 - \beta t)\right]. \quad (2.49)$$

When discrete observations are made at successive positive and negative peaks, the phases of the pendulum are $m\pi/2$ with m an odd integer, and at those peaks

$$x_s^2 \approx \left(\frac{Ct}{2\omega_1}\right)^2, \quad (2.50)$$

while

$$\langle x_n^2 \rangle + \langle x_i^2 \rangle = \left(\frac{\tau_0 u_n}{2\pi}\right)^2 \left[1 - \left(\frac{\beta}{2\omega_1}\right)^2 (1 - \beta t)\right]. \quad (2.51)$$

Consequently the least detectable acceleration $C_{\min}$ is given by

$$C_{\min} \geqslant \frac{2}{t} u_n. \quad (2.52)$$

Since u_n is the rms noise velocity, it will be seen that $C_{\min}$ is of the form

$$C_{\min} \geqslant \frac{\text{rms noise velocity}}{\text{characteristic time}}. \quad (2.53)$$

The characteristic time introduced in eqn (2.53) varies with the type of measurement; we usually omit a numerical factor of order unity. Here the characteristic time is just t, the time of observation. Generally speaking, the characteristic time varies with the type of signal and the way in which the observation is made. We will set out later typical values of characteristic times in different situations.

If $t \gg 2/\beta$, instead of the form of eqn (2.48) we will have

$$x_s^2 \approx \left(\frac{C}{\omega_1 \beta}\right)^2 \sin^2 \omega_1 t \quad (2.54)$$

at the time when the response of the pendulum reaches its maximum amplitude, and eqn (2.51) will be replaced by

$$\langle x_n^2 \rangle + \langle x_i^2 \rangle = \frac{k_b T}{m\omega_0^2}. \quad (2.55)$$

Then with $\sin^2 \omega_1 t = 1$, as before, the least detectable acceleration is given by

$$C_{\min} \geqslant \frac{2}{\tau^*} u_n, \quad (2.56)$$

where $\tau^* = 2/\beta$, the relaxation time, is now the characteristic time. The formula of eqn (2.56) is derived for a resonant force ($p \approx \omega_0$), but in the equilibrium state ($t \gg 2/\beta$) the least detectable signal is independent of the form of the applied force and always has that value (see section 2.5).

(2) *Constant force.* Let the applied force be constant ($p = 0$) and be applied from $t = 0$. Then if $1/\omega_0 < t \ll 2/\beta$, eqn (2.48) can be replaced by

$$x_s^2 = \frac{C^2}{\omega_0^4}\left[1 - (1 - \beta t/2)\left(\cos\omega_1 t - \frac{\beta}{2\omega_1}\sin\omega_1 t\right)\right]^2,\qquad (2.57)$$

while the noise response is given as before by eqn (2.49), and C_{min} can be found from discrete observations. If the observation time t is so chosen that $\cos\omega_1 t = 0$, then

$$x_s^2 = \frac{C^2}{\omega_0^4}\qquad (2.58)$$

and

$$\langle x_n^2\rangle + \langle x_i^2\rangle = \frac{k_b T}{m\omega_0^2}\left[1 - \left(\frac{\beta}{2\omega_1}\right)^2(1 - \beta t)\right].\qquad (2.59)$$

Consequently

$$C_{min} \geqslant \frac{2\pi}{\tau_0}u_n,\qquad (2.60)$$

which is independent of the observation time t. The characteristic time is τ_0, where τ_0 is the free period of the pendulum. In actual experiments, it is usual for the time t to be greater than τ_0, and in consequence the least detectable acceleration in a resonance experiment is smaller than that in an experiment in which a constant force is applied.

2.5 Least detectable signal – continuous measurements

Calculations of the least detectable signal for continuous measurements involve different weighting functions as in eqn (2.26) and will depend on the times t_2 and t_1 which will be chosen to obtain an expected value of the displacement of the pendulum. We first calculate the functional of the noise between the times

$$t_1 = 2n_1\pi/\omega_1\qquad (2.61a)$$

and

$$t_2 = 2n_2\pi/\omega_1,\qquad (2.61b)$$

where n_1 and n_2 are integers.

From eqn (2.13), the functional of $x_i(t)$, is given by

$$\frac{1}{\Delta \tau} \int_{t_1}^{t_2} e^{-\beta t/2} \left[x_0 \left(\cos \omega_1 t + \frac{\beta}{2\omega_1} \sin \omega_1 t \right) + \frac{u_0}{\omega_1} \sin \omega_1 t \right] G(t) \, dt. \quad (2.62)$$

If $G(t) = 1$, the functional evaluates to

$$\frac{1}{\Delta \tau \omega_1} \left[x_0 \left(\sin \omega_1 t + \frac{\beta^2}{4\omega_1^2} \sin \omega_1 t \right) - \frac{u_0}{\omega_1} \left(\cos \omega_1 t + \frac{\beta}{2\omega_1} \sin \omega_1 t \right) \right] e^{-\beta t/2} \Big|_{t_1}^{t_2}.$$

With the above values of t_2 and t_1, and assuming that $\beta t_1, \beta t_2$ are small, the functional becomes

$$\frac{\beta u_0}{2\omega_1^2}.$$

Now by the equipartition theorem the variance of u_0 is u_n^2, and thus the ensemble variance of the functional can be written as

$$\left\langle \left(\frac{1}{\Delta \tau} \int_{t_1}^{t_2} x_i \, dt \right)^2 \right\rangle = \frac{1}{4\pi^2} \frac{\tau_0^2}{\tau^{*2}} \frac{u_n^2}{\omega_1^2}. \quad (2.63)$$

A comparison with eqn (2.41) shows that the expectation of the functional can be much smaller than that of a discrete measurement.

The functional of the noise response, again with $G(t) = 1$, is given by

$$\frac{1}{\Delta \tau} \int_{t_1}^{t_2} x_n(t) G(t) \, dt = \frac{1}{\Delta \tau} \int_{t_1}^{t_2} \left[\frac{1}{\omega_1} \int_0^t F_n(\xi) e^{-\beta(t-\xi)/2} \sin \omega_1(t-\xi) \, d\xi \right] dt. \quad (2.64)$$

It is not necessary to evaluate that integral for it is the ensemble mean square that is required

$$\left\langle \left[\frac{1}{\Delta \tau} \int_{t_1}^{t_2} x_n(t) \, dt \right]^2 \right\rangle,$$

and that may be estimated from the following inequality:

$$\frac{1}{\Delta \tau} \int_{t_1}^{t_2} \langle [x_n(t)]^2 \rangle \, dt \geqslant \left\langle \left[\frac{1}{\Delta \tau} \int_{t_1}^{t_2} x_n(t) \, dt \right]^2 \right\rangle, \quad (2.65)$$

which is a particular case of the Cauchy–Schwarz–Bunyakovskii inequality:

$$\left[\int_a^b f(x) g(x) \, dx \right]^2 \leqslant \int_a^b [f(x)]^2 \, dx \int_a^b [g(x)]^2 \, dx. \quad (2.66)$$

The left side of eqn (2.65) can be calculated from the results of eqns (2.43) and (2.45), when β is very small; we then have

$$\left\langle \left[\frac{1}{\Delta\tau} \int_{t_1}^{t_2} x_n(t)\,dt \right]^2 \right\rangle \leqslant \frac{1}{\Delta\tau} \int_{t_1}^{t_2} \langle [x_n(t)]^2 \rangle \,dt$$

$$\approx \frac{\beta(t_1+t_2)}{2} \frac{k_b T}{m\omega_1^2} + \frac{\beta}{\omega_1^2} \frac{\cos 2\omega_1 t_2 - \cos 2\omega_1 t_1}{t_2 - t_1} \frac{k_b T}{4m\omega_1^2}. \quad (2.67)$$

Thus, finally,

$$\left\langle \left(\frac{1}{\Delta\tau} \int_{t_1}^{t_2} x_s\,dt \right)^2 \right\rangle + \left\langle \left[\frac{1}{\Delta\tau} \int_{t_1}^{t_2} x_n(t)\,dt \right]^2 \right\rangle \leqslant \frac{\beta(t_1+t_2)}{2} \frac{k_b T}{m\omega_1^2} = \frac{2\pi(n_1+n_2)}{\omega_1^3 \tau^*} u_n^2. \quad (2.68)$$

The value of the functional of x_s depends on the period of the applied force and the form of $G(t)$ must be chosen accordingly. We discuss special cases.

(1) *Constant force.* The appropriate functional is the simple average; from eqn (2.23), the functional of x_s, when $G(t) = 1$ and $p = 0$, is given by

$$\frac{1}{\Delta\tau} \int_{t_1}^{t_2} x_s\,dt = \frac{1}{\Delta\tau} \int_{t_1}^{t_2} \frac{C}{\omega_0^2} \left[1 - e^{-\beta t/2} \left(\cos\omega_1 t - \frac{\beta}{2\omega_1} \sin\omega_1 t \right) \right] dt, \quad (2.69)$$

where $\Delta\tau$ is the span over several free periods of the pendulum. For arbitrary values of t_1 and t_2, the result is

$$\frac{1}{\Delta\tau} \int_{t_1}^{t_2} x_s\,dt = \frac{C}{\omega_0^2} - \frac{C}{\Delta\tau\omega_0^2\omega_1} \left(\sin\omega_1 t - \frac{\beta}{2\omega_1} \cos\omega_1 t \right) e^{-\beta t/2} \Big|_{t_1}^{t_2}$$

$$+ \frac{C\beta}{2\Delta\tau\omega_0^2\omega_1^2} \left(\cos\omega_1 t + \frac{\beta}{2\omega_1} \sin\omega_1 t \right) e^{-\beta t/2} \Big|_{t_1}^{t_2}. \quad (2.70)$$

Evidently the best estimate of C is obtained by setting

$$\cos\omega_1 t_1 = \cos\omega_1 t_2 = 1. \quad (2.71)$$

The functional is then

$$\frac{1}{\Delta\tau} \int_{t_1}^{t_2} x_s\,dt = \frac{C}{\omega_0^2} - \frac{C\beta^2}{4\omega_0^2\omega_1^2}, \quad (2.72)$$

which is the expected value of the displacement to be measured.

The values of t_1 and t_2 used in eqns (2.71) and (2.61) are the same.

Therefore the least detectable acceleration can be obtained by substituting eqns (2.72) and (2.68) into (2.39), for the criterion for the least detectable signal, and is

$$C_{\min} \geqslant \omega_0 \left(\frac{k_b T}{m}\right)^{\frac{1}{2}} \left[\frac{\beta(t_1+t_2)}{2}\right]^{\frac{1}{2}} = \frac{2\pi(n_1+n_2)^{\frac{1}{2}}}{(\tau_0 \tau^*)^{\frac{1}{2}}} u_n, \tag{2.73}$$

where n_1 and n_2 are integers (see eqn (2.61)).

The characteristic time, $[\tau_0 \tau^*/(n_1+n_2)]^{\frac{1}{2}}$, is much larger than that in the discrete measurement. A comparison of the results of eqns (2.73) and (2.60) shows that the noise limit for continuous measurements is less than that for discrete measurements by a factor $[\tau^*/(n_1+n_2)\tau_0]^{\frac{1}{2}}$, which is quite significant. If n_1, n_2 are small, the reduction may reach nearly $Q^{\frac{1}{2}}$, where Q is the quality factor of the system. Thus for a torsion pendulum with a high Q a reduction factor of 100 or more may be easily achieved in continuous measurement. However, eqn (2.73) shows that the value of the least detected acceleration increases as the time t_1+t_2 increases. That is because in the non-equilibrium state the noise response increases with time, but the response of the signal remains unchanged since the signal is constant; there is no build up of the amplitude as in the case of resonance.

(2) *Resonance.* In order to obtain the expected value of $x_s(t)$ which is proportional to C, the appropriate weighting function is $G(t) = +1$ in the positive half-cycle of the signal and $G(t) = -1$ in the negative half-cycle. In one cycle, the functional is then

$$\overline{x_s(t)} = \frac{1}{\tau_0}\left(\int_{2\pi n/\omega_1}^{2\pi(n+\frac{1}{2})/\omega_1} x_s \, dt - \int_{2\pi(n+\frac{1}{2})/\omega_1}^{2\pi(n+1)/\omega_1} x_s \, dt\right). \tag{2.74}$$

With

$$x_s(t) = \frac{C}{\omega_1 \beta}(1 - e^{-\beta t/2}) \sin \omega_1 t, \tag{2.75}$$

which is of course the same as eqn (2.25), and $\beta t \ll 1$, the result is

$$(4n+1)\,C/2\omega_1^2.$$

For the purpose of comparing the functional of $x_s(t)$ with that of noise, the span $t_2 - t_1$ must be the same; thus, in the span $t_1 = 2n\pi/\omega_1 = 2n_1\pi/\omega_1$ to $t_2 = 2n_2\pi/\omega_1$, as defined in eqn (2.61), if (n_2-n_1) is not very large, the expected value of the displacement is

$$\frac{1}{\Delta \tau}\int_{t_1}^{t_2} x_s \, dt = (4n_1+1)(n_2-n_1)\frac{C}{\omega_1^2}. \tag{2.76}$$

The least detectable acceleration is now given by

$$C_{\min} \geqslant \frac{(n_1 + n_2)^{\frac{1}{2}}}{(n_2 - n_1)(4n_1 + 1)} \frac{2\pi}{(\tau_0 \tau^*)^{\frac{1}{2}}} u_n. \tag{2.77}$$

Now the characteristic time is $(\tau_0 \tau^*)^{\frac{1}{2}} (4n_1 + 1)(n_2 - n_1)/(n_1 + n_2)^{\frac{1}{2}}$, and if $n_1 \approx n_2$, it reduces to about $(n_1 \tau_0 \tau^*)^{\frac{1}{2}} = (\tau_1 \tau^*)^{\frac{1}{2}}$. In this case, the limit is less than that of discrete measurement (see eqn (2.52)) by a factor $(\tau^*/t)^{\frac{1}{2}}$. The right side of eqn (2.77) is approximately

$$\frac{u_n}{(\tau^* t)^{\frac{1}{2}}}.$$

Comparing the result in eqn (2.77) and that in eqn (2.52), it is found the noise level can be reduced by a factor $(\tau^*/t)^{\frac{1}{2}}$ when continuous measurement is used. This is quite encouraging as it shows that it may be possible to perform significant post-Newtonian laboratory experiments using a torsion pendulum, although at the cost of more complicated data processing.

In sections 2.4 and 2.5 the least detectable forces were estimated for applied forces of special forms. However, when the system reaches equilibrium, there is a limit that is independent of the form of the applied force, as shown in the next section.

2.6 Least detectable force at equilibrium

If the observation time is much longer than the relaxation time $2/\beta$, the linear oscillator will be in an equilibrium state, and the least detectable force is independent of the form of the force. The power spectrum of the thermal driving force, according to the generalized Nyquist theorem, is

$$P(\omega) = 4Hk_b T. \tag{2.78}$$

In the equilibrium state, the displacement response of a torsion pendulum to thermal noise is of Lorentzian form centred at the natural frequency: from eqn (2.5) the power spectrum of $x(t)$ when the forcing term is just thermal noise, is

$$P(\omega) = \frac{4\beta k_b T}{(\omega_0^2 - \omega^2)^2 + \beta^2 \omega^2}. \tag{2.79}$$

The force that acts on the linear oscillator depends on the bandwidth $\Delta\omega$ of the filter constituted by the oscillator and is $(4Hk_b T\Delta\omega)^{\frac{1}{2}}$; the least detectable force must be greater than this force. The effective bandwidth can be estimated by requiring that some specific fraction of the power $P(\omega)$ should fall within the bandwidth.

If the limits of the bandwidth so defined are ω_{max} and ω_{min}, then because of the symmetrical Lorentzian form,

$$P(\omega_{min}) = P(\omega_{max}) = \eta^4 P(\omega_0), \tag{2.80}$$

where η is a dimensionless reliability factor which will usually be less than unity. With the foregoing expression for the power spectrum, the maximum and minimum frequency become

$$\omega_{max} = \omega_0 \left[1 - \frac{2}{Q^2} + 2 \left(\frac{1}{Q^4} - \frac{1}{Q^2} + \frac{1}{Q^2 \eta^4} \right)^{\frac{1}{2}} \right]^{\frac{1}{2}}, \tag{2.81}$$

$$\omega_{min} = \omega_0 \left[1 - \frac{2}{Q^2} + 2 \left(\frac{1}{Q^4} - \frac{1}{Q^2} - \frac{1}{Q^2 \eta^4} \right)^{\frac{1}{2}} \right]^{\frac{1}{2}}. \tag{2.82}$$

If $Q \gg 1$, $\eta^4 \ll 1$, then $\Delta\omega$, which is $\omega_{max} - \omega_{min}$, is given by

$$\Delta\omega \approx \omega_0 \left[\left(1 + \frac{2}{Q\eta^2} \right)^{\frac{1}{2}} - \left(1 - \frac{2}{Q\eta^2} \right)^{\frac{1}{2}} \right] \approx \frac{2\omega_0}{Q\eta^2}, \tag{2.83}$$

which gives the order of the bandwidth. The thermal noise forcing function is then

$$f(t) \approx \frac{2}{\eta} \left(\frac{8\pi m k_b T}{\tau^* Q \tau_0} \right)^{\frac{1}{2}} \tag{2.84}$$

and the minimum detectable acceleration is

$$C_{min} \geqslant \frac{4}{\eta} \frac{u_n}{\tau^*}, \tag{2.85}$$

an expression similar to eqn (2.56) which was obtained in a special case. From eqn (2.85), we know that the least detectable force in an equilibrium state depends only on the mechanical properties of the torsion pendulum; the characteristic time is roughly the relaxation time.

2.7 Characteristic times in the expression for least detectable acceleration

The results of sections 2.3–2.5 are summarized in Table 2.1 (see Chen & Cook, 1990a) which gives the values of the characteristic time in the expression

$$C_{min} \geqslant \frac{\text{rms noise velocity}}{\text{characteristic time}}.$$

Table 2.1. *The characteristic times in different situations*

Measurement methods	Frequency of applied force	Measurement time	Characteristic time
Discrete	$p = \omega_0$	$t \ll \tau^*$	t
		$t \gg \tau^*$	τ^*
	$p = 0$	$t \ll \tau^*$	τ_0
		$t \gg \tau^*$	τ^*
Continuous	$p = \omega_0$	$t \ll \tau^*$	$(t\tau^*)^{\frac{1}{2}}$
		$t \gg \tau^*$	τ^*
	$p = 0$	$t \ll \tau^*$	$(\tau_0 \tau^*/n)^{\frac{1}{2}}$
		$t \gg \tau^*$	τ^*

2.8 Thermal limit on measurements of period

In some experiments a force is measured by the change it produces in the period of a linear harmonic oscillator. The effect of thermal noise upon the period can be treated quite briefly because most of the required formulae have already been developed.

Let the period of the linear oscillator which acts as a force detector be estimated from measurements of the times at which the oscillator passes the 'rest' position; that position will vary with the fluctuation of the displacement of the linear oscillator under the influence of the thermal driving force as previously discussed. If the initial position $x_0 = 0$ is assumed to be the reference for the rest position then the mean square variation of that position is given by eqn (2.46). The instantaneous change of the period is related to the fluctuation of the 'rest' position by the condition

$$a \sin \omega t + (\langle x_n^2 \rangle + \langle x_i^2 \rangle)^{\frac{1}{2}} = a \sin (\omega + \Delta\omega) t, \qquad (2.86)$$

where a is the amplitude of the swing. Hence, taking variances,

$$\langle (\Delta\omega)^2 \rangle = (\langle x_n^2 \rangle + \langle x_i^2 \rangle)/a^2 t^2. \qquad (2.87)$$

In fact the amplitude of swing will decay with time and, taking account of that, by the result of eqn (2.46) we will have

$$\langle \Delta\omega^2 \rangle^{\frac{1}{2}} = \frac{e^{\beta t/2}}{a_0 t} \left\{ \frac{k_b T}{\lambda} \left[1 - \left(\cos \omega_1 t + \frac{\beta}{2\omega_1} \sin \omega_1 t \right)^2 e^{-\beta t} \right] \right\}^{\frac{1}{2}}. \qquad (2.88)$$

It might seem that the fluctuation of the period depends upon the phase

$\omega_1 t$, but that is not so. To determine a period of swing, measurements are made at the time when the oscillator passes the rest position, so that t is an integral multiple of the period τ_0. Therefore eqn (2.88) becomes

$$\langle \Delta \omega^2 \rangle^{\frac{1}{2}} = \frac{e^{\beta t/2}}{a_0 t} \left[\frac{k_b T}{\lambda} (1 - e^{-\beta t}) \right]^{\frac{1}{2}}, \tag{2.89}$$

a result also obtained by A. B. Pippard (1989, personal communication). The uncertainty of the frequency given by eqn (2.89) decreases with time to a minimum value but then increases with time; there is accordingly an optimum measurement time t for a minimum effect of thermal noise given by the equation

$$\beta t \, e^{\beta t} - 2 e^{\beta t} + 2 = 0, \tag{2.90}$$

the solution of which is

$$t \approx 1.5936/\beta = 0.79681 \, \tau^*. \tag{2.91}$$

Thus the least fluctuation in the period due to thermal noise is

$$\langle \Delta \omega^2 \rangle^{\frac{1}{2}} = \frac{0.62}{Q a_0} u_n. \tag{2.92}$$

For a practical example, if $Q = 10^6$, $a_0 = 0.01$ m, $m = 0.05$ kg and $T = 300$ K, this fluctuation is about 10^{-14} s^{-1}.

2.9 Electrical damping of a torsion pendulum

Because the fundamental limit to the sensitivity of linear detectors is set by thermal noise, it is essential to reduce that noise as far as possible. There are two way of reducing the effect of noise: one is to reduce the temperature of the apparatus and the other is to increase the characteristic time of the experiment, that is to say, the time over which the noise response is averaged. Increasing the characteristic time is usually the more effective and cheaper.

   If the main damping of the motion of a pendulum is by the surrounding air, then the Brownian motion and damping are not independent, for the same molecular forces that drive the former are also responsible for the damping. It might be thought therefore that both could be reduced by reducing the air pressure in the apparatus by cooling it to a very low temperature or otherwise, but that is in practice undesirable because the equipment needed to cool the apparatus or to pump it continuously usually produces some mechanical vibration: it is best not to pump or cool during

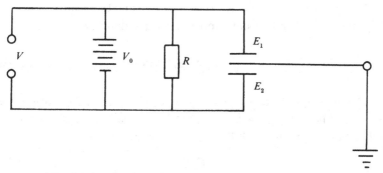

Fig. 2.1 Application of feedback to a torsional pendulum.

the course of observations but to seal the apparatus and let it run passively. There is the further problem that if the damping is too small, the motion of the pendulum is likely to be inconveniently irregular. It is assumed here that the damping is mainly the result of molecular impacts, but if the air pressure is low enough, both damping and noise are the results of friction in the torsion fibre.

Evidently there would be advantage in controlling the effective temperature and the damping independently, as by some form of electrical feedback. There are various instances in which noise has been reduced by feedback, for example acoustic noise and the thermal noise of antiprotons in a storage ring, so that its use with an oscillator such as a torsion balance should be considered. Instruments such as seismometers are commonly provided with feedback to control the period or the damping or to stabilize the rest position, but thermal noise is not significant in such applications. Electrical damping of a galvanometer has been discussed (Ornstein, 1927; Ornstein *et al.*, 1927), as well as that of a bar antenna for the detection of gravitational radiation (Hirakawa *et al.*, 1976).

When electrical feedback is applied to a mechanical oscillator, it contributes to the total thermal noise and it also alters the response to an applied force or acceleration, and the overall signal-to-noise ratio of the oscillator will be the result of the combination of those two factors.

Rather than develop a general theory, which would have to cover very many different schemes of feedback, we illustrate the issues by discussing a rather simple arrangement, shown in Fig. 2.1. Here a voltage V is applied to fixed electrodes, E_1 and E_2 on either side of one of the arms of a torsion pendulum. It is a common way of applying an electrical force. If the feedback is to damp the motion, V will be proportional to the velocity of the pendulum.

Let V_1 and V_2 be the voltages between the beam and the electrodes E_1 and E_2 respectively. The electrostatic force on the pendulum f_e is

$$f_e = \tfrac{1}{2}V_1^2\frac{\mathrm{d}C_1}{\mathrm{d}x} + \tfrac{1}{2}V_2^2\frac{\mathrm{d}C_2}{\mathrm{d}x}, \qquad (2.93)$$

where C_1 and C_2 are the capacitances between the beam and the respective electrodes. When the beam is displaced by a small amount x from its equilibrium position in which the capacitances are equal,

$$C_1 = C_0 + ax, \qquad (2.94)$$
$$C_2 = C_0 - ax, \qquad (2.95)$$

where C_0 and a are constants.

Further

$$V_1 = V + \tfrac{1}{2}V_0, \qquad (2.96)$$
$$V_2 = V - \tfrac{1}{2}V_0, \qquad (2.97)$$

where V_0 is a fixed bias voltage applied between the electrodes.

Equation (2.93) then reads

$$f_e = \tfrac{1}{2}(V_1 - V_2)(V_1 + V_2)a \approx VV_0 a, \qquad (2.98)$$

when x is small.

The motion of the torsion pendulum, therefore, follows the equation

$$m\ddot{x} + H\dot{x} + \lambda x = aV_0 V + n(t). \qquad (2.99)$$

The difference between the charges Q_1 and Q_2 on the plates E_1 and E_2 respectively, when the resistance R is large, is given by

$$Q_e = (V + \tfrac{1}{2}V_0)C_1 - [-(V - \tfrac{1}{2}V_0)C_2] = 2VC_0 + V_0 ax. \qquad (2.100)$$

Changes in Q_e drive current through the resistor, and the voltage V is given by the equation

$$R\frac{\mathrm{d}Q_e}{\mathrm{d}t} + V = E(t), \qquad (2.101)$$

where $E(t)$ is the thermal fluctuation voltage in the resistor R.

On substituting for Q_e from eqn (2.100) we have

$$2RC_0\frac{\mathrm{d}V}{\mathrm{d}t} + aV_0 R\frac{\mathrm{d}x}{\mathrm{d}t} + V = E(t). \qquad (2.102)$$

On eliminating V and $\mathrm{d}V/\mathrm{d}t$ between eqns (2.99) and (2.102), the equation satisfied by x, the displacement of the end of the torsion beam, is found to be

$$2mRC_0\frac{\mathrm{d}^3x}{\mathrm{d}t^3}+(2HRC_0+m)\frac{\mathrm{d}^2x}{\mathrm{d}t^2}+(2\lambda RC_0+H+a^2V_0^2R)\frac{\mathrm{d}x}{\mathrm{d}t}+\lambda x$$

$$= aV_0E(t)+2RC_0\frac{\mathrm{d}n(t)}{\mathrm{d}t}+n(t). \quad (2.103)$$

It is often the case that the time constraint RC_0 is very small; then the third-order equation reduces to one of second order:

$$m\frac{\mathrm{d}^2x}{\mathrm{d}t^2}+(H+a^2V_0^2R)\frac{\mathrm{d}x}{\mathrm{d}t}+\lambda x = aV_0E(t)+n(t). \quad (2.104)$$

A similar equation of second order was derived by Ornstein *et al.* (1927) in their study of the behaviour of a galvanometer; in that case it was the inductance of the coil that was neglected in order to obtain the equation of second order. In a similar way the equation of motion of a torsion pendulum with electrostatic feedback is reduced to one of second order if the capacitance of the electrodes is supposed to be very small. That is in fact usually so – in the apparatus of Chen, Cook & Metherell (1984) the capacitance of the feedback electrodes was about 0.5 pF, about that of a 1 m length of coaxial cable.

The mean square fluctuation of electrical origin may be denoted by $\langle x_e^2\rangle$ and that of mechanical origin by $\langle x_m^2\rangle$ and they are to be calculated from

$$\langle x^2\rangle = \int_0^\infty |W(f)|^2 G(f)\,\mathrm{d}f, \quad (2.105)$$

where $G(f)$ is the power spectrum of the thermal fluctuation force and $W(f)$ is the complex response function of the equation of motion; if the equation of motion is

$$m\ddot{x}+\beta\dot{x}+\lambda x = 0 \quad (2.106)$$

then

$$W(f) = \frac{1}{\lambda-m\omega^2+i\beta\omega}. \quad (2.107)$$

In the present case,

$$\beta = H+a^2V_0^2R. \quad (2.108)$$

The fluctuations of mechanical origin and electrical origin are independent because the equation of motion is linear. The spectrum of the mechanical thermal force $n(t)$ is, by the generalized Nyquist theorem,

$$G_m(f) = 4Hk_b T_m, \quad (2.109)$$

where T_m is the temperature of the mechanical system.

Thus

$$\langle x_m^2 \rangle = \int_0^\infty \left| \frac{1}{\lambda - m\omega^2 + i\beta\omega} \right|^2 8\pi H k_b T_m \, d\omega = \frac{k_b H T_m}{(H + a^2 V_0^2 R)\lambda}. \quad (2.110)$$

Similarly the power spectrum of the electrical fluctuations is

$$G_e(t) = 4k_b R T_r, \quad (2.111)$$

where T_r is the temperature of the resistor.

The mean square fluctuation of electrical origin is then

$$\langle x_e^2 \rangle = \int_0^\infty \left| \frac{aV_0}{\lambda - m\omega^2 + i\beta} \right|^2 8\pi R k_b T_r \, d\omega = \frac{k_b a^2 V_0^2 R T_r}{(H + a^2 V_0^2 R)\lambda}. \quad (2.112)$$

When the electrical and mechanical temperatures are the same, the total fluctuation is

$$\langle x^2 \rangle = \langle x_m^2 \rangle + \langle x_e^2 \rangle$$

or

$$\langle x^2 \rangle = \frac{k_b T}{\lambda}, \quad (2.113)$$

which is the value to be expected according to the equipartition theorem. Thus electrical feedback from circuits at the same temperature as the mechanical system does not change the overall thermal fluctuation.

It follows that the ratio of electrical damping by feedback to the damping by mechanical friction will not alter the noise fluctuations and so electrical feedback may be applied if it is required for other reasons without increasing the noise response, but on the other hand feedback from a circuit at the same temperature as the mechanical system cannot reduce the overall noise fluctuation. We now go on to discuss the effect of cooling the electrical circuits and specifically the resistor R in the circuit of Fig. 2.1.

2.10 Partial cooling of a torsion pendulum

The fluctuation of a system with electrical feedback depends on both the mechanical damping and temperature and the electrical resistor and its temperature and it may be seen from eqns (2.110) and (2.112) that the total mean square fluctuation when the temperatures are unequal is

$$\langle x^2 \rangle = \frac{H}{(H + a^2 V_0^2 R)\lambda} k_b T_m + \frac{a^2 V_0^2 R}{(H + a^2 V_0^2 R)\lambda} k_b T_r. \quad (2.114)$$

An effective temperature may then be defined as

$$T_e = T_r + \frac{H}{H + a^2 V_0^2 R}(T_m - T_r).$$ (2.115)

Obviously if T_m and T_r are the same, the effective temperature is also equal to them as has already been seen.

If T_r is less than T_m then T_e is also less than T_m, and if T_r is cooled to a temperature much less than T_m then

$$T_e \approx T_m \left(\frac{H}{H + a^2 V_0^2 R} \right).$$ (2.116)

Thus the ratio of T_e to T_m can be quite small if $a^2 V_0^2 R$ is much greater than H, that is, if the electrical damping is much greater than the mechanical damping, and then the cooling of the electrical circuit will be very effective.

However, reduction of the noise temperature does not necessarily mean that the signal-to-noise ratio is improved, for the increased electrical damping also reduces the signal. Suppose we consider the case where the measurement time is longer than the relaxation time, as it might be if the electrical damping is great. The characteristic time to be used in estimating the signal-to-noise ratio is then the relaxation time and that is now

$$\frac{2m}{H + a^2 V_0^2 R}$$

instead of $2m/H$.

Hence the value of C_{min} is increased in the ratio $[(H + a^2 V_0^2 R)/H]^{\frac{1}{2}}$, and since the noise is proportional to the square root of the effective temperature, we can see that the signal-to-noise ratio is the same as for purely mechanical damping.

A similar conclusion was reached by Hirakawa *et al.* (1976) but they point out that there may none the less be an advantage in cooling the feedback system. If the noise fluctuations are reduced, then the detector and its associated amplifiers and recorders need not have so large a dynamic range and fewer digits need be retained in the subsequent signal processing.

3

External sources of noise and design of experiments

3.1 Introduction

Thermal noise is unavoidable and sets the fundamental limit to the detectability of the response of an oscillator to any gravitational effect, but it is not the only disturbance to which an oscillator may be subject. Other forces may act on the mass of a torsion pendulum if it is subject to electric or magnetic or extraneous gravitational fields. The point of support of a torsion pendulum or other mechanical oscillator may be disturbed by ground motion. Ground motion is predominantly translational and so might be thought not to affect a torsion pendulum to a first approximation. However, all practical oscillators have parasitic modes of oscillation besides the dominant one, and although in linear theory normal modes are independent, in real non-linear systems modes are coupled. Thus, even if in theory seismic ground motion had no component of rotation about a vertical axis, none the less there would be some coupling between the primary rotational mode of a torsion pendulum and its oscillations in a vertical plane. In practice, therefore, any disturbance of a mechanical oscillator may masquerade as a response to a gravitational signal.

External sources of noise can be avoided with proper design of experiments. In this chapter we shall discuss both the sources of external disturbance and also the ways in which oscillators of different design respond.

3.2 Ground disturbance

Sources of ground noise

We begin with a discussion of seismic motions that move the point of support of a pendulum. There are three main sources of ground noise: man-made, meteorological and seismic. Generally speaking, man-made

34

sources (such as factories, traffic, trains, airplanes, and the movement of observers) and meteorological sources (such as atmospheric turbulence and winds on buildings and trees) have a high frequency range of 1–100 Hz and generate elastic surface waves in the ground that die away rapidly with depth. If the site of an experiment is far from the sources and also underground, the amplitude of ground noise may be as low as 10^{-7} cm. Noise from a seismic source is much less easily avoided, and may limit the precision of gravitational experiments in the laboratory.

The Earth, as an elastic body, has normal modes of free oscillation that are excited by earthquakes. The modes are of two classes: spheroidal motion, contracting and expanding along the radial direction; and toroidal motion, a relative rotation of parts of the Earth. A record of the ground motion corresponding to the free oscillations has a random appearance but may be resolved into thousands of modes of periods from 54 min downwards and of random phase. Seismic noise in the frequency band of 0.01–1 Hz is caused by storms and waves in the ocean. The amplitude varies with the seasons and the weather, and in bad weather may reach 10^{-2} cm in unfavourable sites. Delicate experiments on gravitation should thus be performed far from oceans and in clear weather if possible.

There are also much slower Earth movements caused by atmospheric loading, ocean tides, orogeny and solar heating that should not affect laboratory experiments. Table 3.1 gives a summary of seismic sources of ground noise.

Tilting of the ground is another source of noise. There are various reasons for the change of the direction of the surface of the ground; the main contribution to the tilting is due to solar radiation which causes thermal deformation of the ground surface. The amplitude of the tilt can be of the order of several seconds of arc.

Response of a linear oscillator to ground noise

Let the point of support of a linear mechanical oscillator move a small amount $\delta(t)$ due to the motion of the ground (it may be a translation or a rotation). It is also assumed that the movement is directly coupled to the motion of the oscillator (for example, translational motion can directly couple to a simple pendulum, rotational motion can directly couple to a torsion balance, etc.). The equation of motion is then

$$\frac{\mathrm{d}^2 x}{\mathrm{d}t^2} + \beta \frac{\mathrm{d}x}{\mathrm{d}t} + \omega_0^2 (x - \delta) = 0. \tag{3.1}$$

Table 3.1. *Frequency bands of seismic ground noise*

Frequency (Hz)	Sources
$1-10^2$	Man-made
$10^{-2}-1$	Storm and waves in the ocean
$10^{-4}-10^{-2}$	Normal oscillation of the Earth
$10^{-5}-10^{-4}$	Tides
$0-10^{-5}$	Orogeny, temperature, etc.

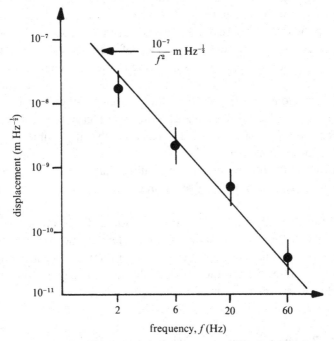

Fig. 3.1 Representative spectrum of ground motion.

Note that the ground motion δ does not enter the damping term. In laboratory gravitational experiments damping can usually be made to depend on the velocity relative to a fixed reference; for example, the eddy current induced by a magnet fixed to the point of support may be used to damp the motion of a pendulum and this point of support is usually very well isolated from the ground motion; the behaviour of a single-base passive vibration insulator with damping relative to a fixed reference is similar (see eqn (3.18) and Fig. 3.5)).

In general the function $\delta(t)$ will be unknown but its spectrum will be known from some survey of the motion of the ground. It has been found (Berger & Levine, 1974; Fix, 1972; J. Hough, 1989, personal communication) that the rms amplitude of the motion of the ground in the frequency band 10^{-8}–10^2 Hz shows an approximately inverse square dependence on the frequency. That is,

$$\Delta(\omega) = A\left(\frac{\text{Hz}}{\omega}\right)^2,\tag{3.2}$$

where $\Delta(\omega)$ is the rms spectral density of the ground displacements. The value of the constant A varies from place to place, a typical figure is from 10^{-8} to 10^{-5} cm $\text{Hz}^{-\frac{1}{2}}$. Fig. 3.1 gives a typical spectrum of ground motion measured in a quiet place in Scotland by J. Hough (1989, personal communication).

In the following we restrict the study to the steady state after transients have died away. Let $X(\omega)$ be the spectral density of the displacement $x(t)$ given by the steady state; from eqn (3.1), it follows immediately that

$$X(\omega) = \frac{\omega_0^2\,\Delta(\omega)}{[(\omega_0^2-\omega^2)^2+\beta^2\omega^2]^{\frac{1}{2}}}.\tag{3.3}$$

Using the value of $\Delta(\omega)$ given by eqn (3.2), the explicit expression for the spectral density of $x(t)$ is

$$X(\omega) = \frac{A\omega_0^2}{[(\omega_0^2-\omega^2)^2+\beta^2\omega^2]^{\frac{1}{2}}}\left(\frac{\text{Hz}}{\omega}\right)^2.\tag{3.4}$$

Fig. 3.2 shows the variation of spectral density of displacement with frequency and the resonant peak at

$$\omega = \omega_r = (\omega_0^2-\beta^2/2)^{\frac{1}{2}}.\tag{3.5}$$

In the high frequency range, $X(\omega)$ is inversely proportional to ω^4 but in the low frequency range it is considerably greater, implying that the effect of ground noise at low frequencies is usually very troublesome in gravitational experiments. That is a typical response.

Generally an oscillator may have more than one mode of oscillation. For example, a torsion pendulum may have rotational motions, swing motions or rocking motions, any of which may couple to the rotational motion of attracted masses hung from the end of the arm. However, only one mode is used to detect the gravitational force (usually the mode most affected by the gravitational interaction or the mode with the lowest frequency). If the

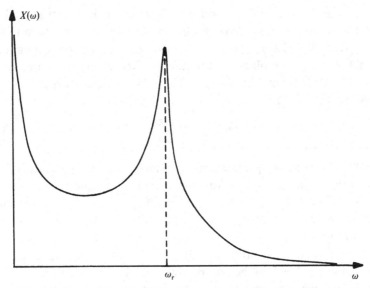

Fig. 3.2 Spectral density of displacement caused by ground motion.

oscillator were linear and perfect, there would be no coupling between different modes. Unfortunately, a real oscillator is not strictly linear and perfect, and different modes of vibration may be coupled by non-linear effects and other imperfections of the oscillator. In this sense, the study is not complete, and ground noise may give more serious effects than obtained above.

Before discussing how the influence of ground noise can be reduced, we first look at how the ground noise in a typical quiet location limits the precision of an experiment, and that will help us to appreciate the need for isolation from ground motion.

Limit set by ground noise on least detectable acceleration

A general study is not useful because the effects of ground noise depend both upon the location of the experiment and also on the type of oscillator. Thus the effect of ground motion on the torsion balance is complicated for it is difficult to predict the behaviour of the test masses (which may be hung at some distance below the beam of the balance) relative to the beam. The effect on the torsion balance can be discussed in terms of four components of the motion of the ground: rotations about a vertical axis, and horizontal, vertical and tilting vibrations. To avoid complications, we assume that the test masses are fixed at the ends of the beams of the framework in the same horizontal plane, so that the horizontal, vertical and tilting motions have only second-order effects, and they will produce a torque only through the imperfections and the non-linearities of the torsion pendulum system or through the spatially varying external force field as the balance moves as a whole. The rotational vibration of the laboratory will not necessarily exert a direct torque on the balance for that will depend on the location of the axis of the rotation. If the axis does not pass through the point of suspension of the fibre, the effect of a rotation can be resolved into a translation and a pure rotational motion, the latter being a rotation about an axis passing through the point of suspension. Assuming that the power spectrum of the rotational vibration is of the same order and same form as in eqn (3.2), we can estimate the limit of an acceleration detectable by a torsion balance in the presence of ground noise. Consider just ground motion with nearly the same frequency as that of the torsion balance. The acceleration of the torsion balance subject to ground noise with a spectral density $\Delta(\omega)_0$ is roughly

$$a \approx x_r \omega_0^2 = (\Delta(\omega) \Delta f)^{\frac{1}{2}} \omega_0^2, \tag{3.6}$$

where $\Delta(\omega)$ has been defined in eqn (3.2) and Δf is the frequency bandwidth of the response of the system. In eqn (3.2), the value of A can be replaced by a typical value of 10^{-7} cm Hz$^{-\frac{1}{2}}$. Then the limiting acceleration by the ground motion is

$$a = \frac{10^{-7}}{f_0^2} (\Delta f)^{\frac{1}{2}} \omega_0^2 \text{ cm Hz}^{-\frac{1}{2}}, \tag{3.7}$$

where $f_0, \omega 0$ are the frequency and angular frequency of the torsion balance. It is not possible to calculate a frequency response as we did in the study of thermal noise, since the driving force of ground noise, unlike the thermal driving force, depends on frequency (it is 'coloured'). However, taking Δf to be $1/Q\tau_0$ and putting $f_0 = 10^{-4}$ and $Q = 6 \times 10^6$, the limiting acceleration is about 10^{-17} m s^{-2}. Comparing this figure with that of thermal noise, we see that the effect of ground noise can be considerable. It is true that the above figure is a very rough estimate but while a value of $A = 10^{-7}$ cm Hz$^{-\frac{1}{2}}$ may be an overestimate, at the same time, $\Delta f = 1/Q\tau_0$ may be an underestimate. The effects of ground motion can be greatly reduced by mechanical isolating filters, and it is essential to use them in gravitational experiments.

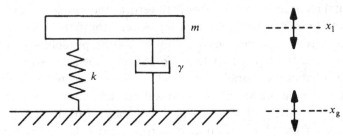

Fig. 3.3 Single-base passive vibration isolator with damping relative to the ground.

Passive isolation of ground motion

Passive systems are often used to isolate experiments from ground noise and we now discuss several such schemes. They may be distinguished as single-base or multi-base systems and the damping may be relative to the ground or to a vibration-free reference, thereby determining whether the ground motion enters the damping term or not.

(1) *Single-base passive vibration insulator with damping relative to the ground.* This is the scheme most commonly used; it is shown in Fig. 3.3. It is typically constructed of a heavy base supported on the ground by a series of springs with a spring constant k and damped by a viscous resistor characterized by a damping factor 2γ. Consider a single harmonic component of the ground noise given by

$$x_g = x_{g0} \cos \omega t, \qquad (3.8)$$

where x_{g0} is a constant. (The real vibration of the ground will have a spectrum of components.)

If x_1 is the motion of the base, the equation of the motion is

$$m\frac{d^2x_1}{dt^2} + 2\gamma\frac{d}{dt}(x_1 - x_g) + k(x_1 - x_g) = 0, \qquad (3.9)$$

or

$$\frac{d^2x_1}{dt^2} + \beta\frac{d}{dt}(x_1 - x_g) + \omega_1^2(x_1 - x_g) = 0, \qquad (3.10)$$

where

$$\beta = \frac{2\gamma}{m} \qquad (3.11)$$

and

$$\omega_1^2 = \frac{k}{m}. \qquad (3.12)$$

Here we note that the displacement of the ground x_g enters the damping term showing that the damping is relative to the ground. If $x_1 - x_g$ is put equal to y, the above equation becomes

$$\frac{d^2 y}{dt^2} + \beta \frac{dy}{dt} + \omega_1^2 y = x_{g0} \omega^2 \cos \omega t. \tag{3.13}$$

The solution is

$$y = \frac{x_{g0} \omega^2 (\omega_1^2 - \omega^2)}{(\omega_1^2 - \omega^2)^2 + \beta^2 \omega^2} \cos \omega t + \frac{x_{g0} \beta \omega^3}{(\omega_1^2 - \omega^2)^2 + \beta^2 \omega^2} \sin \omega t. \tag{3.14}$$

In terms of the variable x, the motion of the base is

$$x_1 = \left[\frac{x_{g0} \omega^2 (\omega_1^2 - \omega^2)}{(\omega_1^2 - \omega^2)^2 + \beta^2 \omega^2} + x_{g0} \right] \cos \omega t + \frac{x_{g0} \beta \omega^3}{(\omega_1^2 - \omega^2)^2 + \beta^2 \omega^2} \sin \omega t. \tag{3.15}$$

The ratio of the amplitude of the motion x_1 (x_{10}) to that of x_g (x_{g0}) is consequently

$$\frac{x_{10}}{x_{g0}} = \left[\frac{\omega_1^4 + \beta^2 \omega^2}{(\omega_1^2 - \omega^2)^2 + \beta^2 \omega^2} \right]^{\frac{1}{2}}. \tag{3.16}$$

The dependence of x_{10}/x_{g0} on frequency for this single-base passive insulator is shown in Fig. 3.4.

We see from the above expression that the isolator is effective only when

$$\omega > 2^{\frac{1}{2}} \omega_1. \tag{3.17}$$

To achieve good isolation, the natural frequency ω_0 of the isolator must be much lower than the frequency of the oscillator, that is, the mass m should be heavy and the spring constant should be small.

(2) *Single-base passive vibration insulator with damping relative to a fixed reference.* It is easily seen that a disadvantage of the insulator with damping relative to the ground is that while it brings the amplitude of the resonance peak down, it greatly increases the coupling between the base and floor.

That coupling can be reduced by referring the damping to a point independent of the ground, in effect a second vibration isolator. This, although it may be more complicated and expensive, is very effective, as will now be discussed. The scheme is shown in Fig. 3.5.

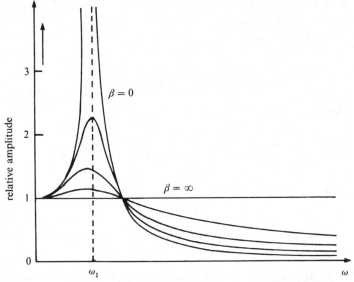

Fig. 3.4 Frequency response of a single-base passive vibration isolator with damping relative to the ground (ω = frequency).

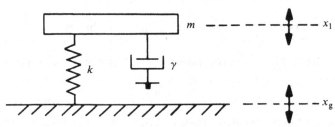

Fig. 3.5 Passive vibration isolator with damping relative to vibration-free reference.

Using the same symbols, eqn (3.13) is now replaced by

$$\frac{\mathrm{d}^2 x_1}{\mathrm{d}t^2} + \beta \frac{\mathrm{d}x_1}{\mathrm{d}t} + \omega_1^2 x_1 = \omega^2 x_{g0} \cos \omega t. \tag{3.18}$$

Then, as before, the ratio of the amplitudes of x_1 and x_g can be expressed by

$$\frac{x_{10}}{x_{g0}} = \left[\frac{\omega_1^4}{(\omega_1^2 - \omega^2)^2 + \beta^2 \omega^2} \right]^{\frac{1}{2}}. \tag{3.19}$$

The comparison of this result with that of eqn (3.16) shows that this kind

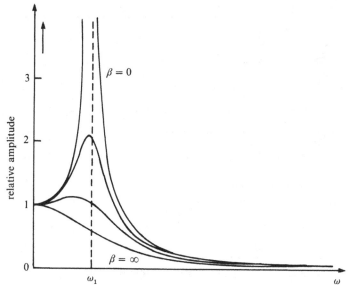

Fig. 3.6 Frequency response of a passive vibration isolator with damping relative to vibration-free reference (ω = frequency).

of isolator may give very effective isolation from ground noise because the condition of eqn (3.17) for the ratio x_{10}/x_{g0} to be less than 1, is replaced by

$$\omega > (2\omega_1^2 - \beta^2)^{\frac{1}{2}}. \tag{3.20}$$

That provides another degree of freedom in designing a passive isolation system, for by a proper choice of β, effective isolation can be obtained at a much lower frequency. The frequency characteristic is shown in Fig. 3.6 and shows that the greater the damping factor, the better the isolation.

In practice of course there is no such thing as a fixed point entirely free of disturbance (at least the Brownian motion cannot be neglected) yet the above comparison indicates that if the damping reference can itself be well isolated from the ground the system as a whole will have an improved performance. The reason is obvious: damping couples the experimental apparatus to the point of reference of the damping.

(3) *Multi-base passive vibration isolator.* An approximation to a system referred to a truly quiet point is given by a number of passive isolators in series. Fig. 3.7 is a schematic diagram for an isolator of n similar elements. For simplicity, the dampings are all assumed to be relative to the respective bases, but we know that if they are relative to a vibration-free reference, the result can be further improved.

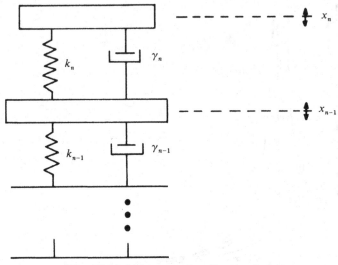

Fig. 3.7 Multi-base passive vibration isolator.

The motion of the multi-base isolator follows the set of equations:

$$\frac{d^2x_n}{dt^2} + \beta_n \frac{d}{dt}(x_n - x_{n-1}) + \omega_n^2(x_n - x_{n-1}) = 0, \tag{3.21}$$

. . . .

$$\frac{d^2x_2}{dt^2} + \beta_2 \frac{d}{dt}(x_2 - x_1) + \omega_2^2(x_2 - x_1) = 0, \tag{3.22}$$

$$\frac{d^2x_1}{dt^2} + \beta_1 \frac{d}{dt}(x_1 - x_g) + \omega_1^2(x_1 - x_g) = 0, \tag{3.23}$$

where $\omega_n^2 = k_n/m_n$ and $\beta_n = 2\gamma_n/m_n$.

The solution is

$$\frac{x_{n0}}{x_{g0}} = \Gamma_n \Gamma_{n-1} \ldots \Gamma_2 \Gamma_1, \tag{3.24}$$

where

$$\Gamma_n = \left[\frac{\omega_n^4 + \beta_n^2 \omega^2}{(\omega_n^2 - \omega^2)^2 + \beta_n^2 \omega^2} \right]^{\frac{1}{2}}. \tag{3.25}$$

The overall effect of a multi-base isolator is simply the product of the factors Γ_n of the individual single-base isolators.

The multi-stage isolator has the advantage that in the range of frequencies for which the individual Γ factors are less than 1, the overall response is much less than that of a single-stage isolator. However, the cut-off frequency at which $\Gamma = 1$ is not altered, and an isolator of a number of identical stages will have the same cut-off frequency as a single stage. In addition each stage will also be subject to Brownian motion so that the disturbance due to the Nyquist force will increase with the number of bases since each separate base will contribute its own degree of freedom to the Brownian motion.

It is then naturally a serious question whether such isolators are useful in gravitational experiments where oscillators used as the gravitational detectors usually have very low characteristic frequencies – a torsion pendulum may have a period of several hours. In fact, even if the cut-off frequency is above the characteristic frequency of the oscillator it is useful to attenuate the higher frequency vibrations because they are always detrimental on account of the non-linearity of the oscillator, while relative motions of parts of the apparatus may produce false signals. Vibration isolators have often been used in gravitational experiments in the laboratory, and a large multi-stage isolator has been developed for a detector of gravitational waves (see Lindsay, Saulson & Weiss, 1983).

There is another effect of the vibration of the ground which cannot be isolated by any means, namely the gravitational effect of the ground motion. Any mechanical vibration of the ground is associated with mass motions that may produce changes in gravitational forces on test masses and therefore some extraneous gravitational effect on a detector. For example, the free oscillations of the Earth give rise to changes in gravity as well as displacements at the surface, and the effect, though small, should be considered in very precise experiments (see section 3.7).

3.3 Magnetic effects

A detector of gravitational forces, such as a torsion pendulum, may be subject to magnetic fields of local and of global origin. A field that produces a constant force or torque will not influence the detection of a gravitational effect, but any change will masquerade as a gravitational effect. Fields of global origin, the Earth's field particularly, change with time: the Earth's field shows a daily variation (Chapman & Bartels, 1940). Local fields are changed by alterations in the disposition of nearby magnetic materials.

Consider the effect of the daily variation of the Earth's field upon a mechanical oscillator. If the structure of the oscillator has a permanent moment, then it will in general be subject to a force or torque proportional to the Earth's field that will therefore have a daily component. If, however, the oscillator has no permanent moment but does have a magnetic permeability, then there will be a force or torque proportional to the square of the field and so there will be a component with a period of twelve hours. The following example gives an idea of how an external magnetic field may affect the torsion balance. In the solar torsion balance experiments to test the weak principle of equivalence (see section 4.8), suppose the torsion frame to be oriented in a symmetrical way along the north–south direction. The expected torque caused by the violation of the weak principle of equivalence is $M_i \Omega^2 \eta L D$, where M_i is the inertial mass of the test body, Ω the angular speed of the Earth in its orbit, η the Eötvös coefficient (section 4.8), L the length of the arm of the balance, and D the distance between the Earth and the Sun, while the magnetic torque due to a dipole $\boldsymbol{M}_m$ in a field $\boldsymbol{B}$ is

$$\tau_m = \boldsymbol{M}_b \times \boldsymbol{B}. \tag{3.26}$$

The ratio of the two torques is of order

$$R = \frac{M_b B}{M_i \Omega^2 \eta L D}, \tag{3.27}$$

and with the condition that R should be less than 10^{-2}, the magnetic dipole of the torsion balance must be smaller than

$$\frac{10^{-2} M_i \Omega^2 \eta L D}{B}. \tag{3.28}$$

If B is taken to be the amplitude of the daily change of the magnetic field of the Earth, namely about 5×10^{-8} T, then the magnetic dipole of the beam must be smaller than 6×10^{-19} J T^{-1}, if the experiment is designed to detect an anomaly with $\eta = 10^{-11}$. (The length of the beam has been taken to be 5 cm and the mass 10 g.)

Local fields may arise either from masses that are part of the experimental apparatus or from those that happen to be in the neighbourhood. The magnetic effect on a gravitational experiment is not only directly on the linear oscillator itself; the magnetic influence of gravitational attracting bodies near the oscillator may be important, especially if the material of the oscillator has been contaminated magnetically. Masses that are involved in gravitational experiments have to be moved to produce the desired gravitational effect and so there is no simple way of separating gravitational and magnetic effects.

The first and most important way of reducing magnetic effects is to ensure that the mechanical oscillator itself is constructed of non-magnetic materials using non-magnetic tools. Copper, gold and mercury, which are quite often used in the gravitational experiments for parts of apparatus, are diamagnetic. They are magnetized along the direction opposite to that of the external magnetic field and have a negative permeability which is usually about 10^{-6}. Another group of commonly used materials includes aluminium and platinum. They are paramagnetic and are magnetized along the same direction as that of the external magnetic field and have a positive permeability which is usually about 10^{-5}.

Iron, nickel, cobalt and certain of their alloys are ferromagnetic with permeabilities that can reach 10^4 or greater. They may also have permanent magnetic moments. The magnetic properties are not constant and depend upon the external field. Ferromagnetic material must be strictly avoided in gravitational experiments. It is in practice very difficult to obtain any metal completely free of ferromagnetic impurities – brass in particular is likely to have them, and a few parts in a million of such impurities could possibly cause great trouble. Not only must the composition of the components be properly chosen but precautions must also be taken to prevent even very small amounts of iron or steel being left on the surface after machining or other operations. The parts of the oscillator must be thoroughly cleaned before assembly.

Not only must apparatus be kept as free as possible of magnetic materials, but the magnetic field to which a mechanical oscillator is subject must also be reduced as much as possible. The steady uniform component of the field can be reduced almost to zero by a field of the opposite sign generated by Helmholtz coils and controlled by a servo-system, but fields of local origin that are not spatially uniform cannot be eliminated in that way. Thus, besides making the mechanical system as far as possible non-magnetic and annulling the field around it as far as possible, it should also be shielded by being placed in a container of high permeability.

Gravitational experiments are usually performed in the weak magnetic field of the Earth. Mumetal, a nickel–iron alloy, has an extremely high permeability in weak fields and makes an efficient shield in those circumstances. The effectiveness of shielding depends on the permeability of the mumetal used and the dimensions. A shielding factor can be defined as the ratio of the magnetic field strengths outside and inside the shielding enclosure:

$$S = \frac{H_{\text{out}}}{H_{\text{in}}}. \tag{3.29}$$

The shielding factor for an infinitely long hollow cylinder in a static field H_{out} is

$$S = \frac{\mu}{2}\frac{t}{d},$$ (3.30)

where μ is the permeability of the metal and t and d are the thickness and diameter of the cylinder.

Obviously, in practice, it is impossible to use an infinite cylinder, but if the distance to the end surface is considerably greater than the diameter, the shielding factor of a finite cylinder can be approximated very well by the formula (3.30). For example, if $\mu = 10^4$ and $t/d = 0.1$, a factor of 500 can be obtained.

Shielding factors of 10^4 or better can be achieved with multi-layer shielding, or superconducting shields may be used (see Phillips, 1965b). While a shielding factor of a thousand or a few thousand is usually sufficient for most purposes in gravitational experiments, a multi-layer mumetal shield will occupy more space around the oscillators, which may lead to an increased distance between the attracted and the attracting mass, which in turn will reduce the gravitational force. A good balance between adequate magnetic shielding and the greatest possible gravitational force must therefore be struck.

3.4 Electrostatic forces

Whenever apparatus is operated at a low enough pressure that the conductivity of the air is very low, electrostatic charges can build up on non-conducting parts of the equipment and produce forces that may give highly erratic results in delicate measurements.

The force f between two charges Q_1 and Q_2, in vacuum, separated by a distance r is

$$f = \frac{Q_1 Q_2}{4\pi\varepsilon_0 r^2},$$ (3.31)

where ε_0 is the permittivity of vacuum ($\varepsilon = 8.85 \times 10^{-12}$ F m^{-1}).

If the parts considered (for example, the test body, the vacuum chamber housing the linear oscillator or any moving part of the oscillator) are made of electrically insulating materials, such as glass, quartz or plastic, they may become electrically charged. Unfortunately, it is impossible to know how the charges are distributed over the bodies and so there is little point in calculating representative forces. The charges and voltages that are generated depend greatly on conditions, but, in general, the better the vacuum the more troublesome electrostatic charging is likely to be.

Electrostatic forces may also arise from potential differences between metals if, for example, a battery is connected. The effect can be calculated from Coulomb's law and Gauss's theorem:

$$\frac{1}{\varepsilon_0}\sum Q = \int_S E \cdot \mathrm{d}S, \tag{3.32}$$

where $\sum Q$ is the total charge inside the surface S of the body considered and E is the electric field strength, effectively the gradient of the voltage differences between the different parts considered.

Electrostatic forces change with time and it is essential to shield all moving parts from them. The following precautions should be taken.

(1) Components should be electrically conducting, or coated with conducting material.
(2) The vacuum chamber can conveniently be made an electric shield. If it is made of glass, a metal wire net or some conducting foil wrapping on the chamber will be sufficient.
(3) The moving parts of the linear oscillator and the vacuum chamber should have the same electrical potential or be earthed at the same point. There is, of course, no reason to earth the vacuum chamber for the purpose of electrostatic shielding because there is no electric field inside a closed conductor. But, to avoid the possible potential difference between the chamber and the moving parts, it is better to earth them together if it is convenient.
(4) The other approach is to irradiate the residual gas in the vacuum chamber with electrons from a strong radioactive source such as used to discharge static electricity on textile machinery.

There may be disadvantages to all these methods in any particular experiment and no general recommendations can be given – trial and error is usually the only way of dealing with static electricity and the purpose of these remarks can be little more than to alert readers to the existence of the problem.

3.5 Miscellaneous thermal effects

As with many delicate experiments, the effects of temperature changes are often the most troublesome in gravitational experiments. Variations of temperature not only couple directly to the mechanical movement of the linear oscillator, for example by causing a movement of a support, but also affect nearly every part of a system: the optical lever, the electronic circuits, the recording system, the power supply, even the change in the form and disposition of nearby masses. The subject is very wide, but in the following we concentrate on discussing how a torsion balance is influenced by direct effects of a change of temperature. It is very important to stabilize the temperature of apparatus and reduce gradients of temperature as much as possible, and in addition to consider carefully the design of experiments and the choice of materials to minimize thermal effects.

We first consider the influence of temperature on contact potentials and on the behaviour of the fibre of a torsion balance.

Temperature dependence of contact potential

Electrodes are used in some gravitational experiments to apply feedback, or to measure displacements, and are usually placed very close to moving parts, and may also be made of different metals. Again the wall of a vacuum chamber may be very close to moving parts in order that the latter may be as close as possible to external attracting masses. Contact potentials may occur in those situations.

Contact potential differences between dissimilar metals exist in a vacuum even if the specimens have been annealed and the effects of strain have been removed. They are usually about 0.5–1 V, provided there is no obvious moisture on the surfaces of the metals, and vary with temperature by about 10^{-3}–10^{-2} V deg^{-1}. The effects are particularly strong if the test masses and the electrodes are prepared in air, because the surfaces of the metals will be oxidized and metallic oxides usually possess very high thermoelectric powers. Even if the temperature coefficient of the contact potential is relatively small, it may be strong enough to produce a movement capable of disturbing the measurement. Test masses and electrodes should therefore be prepared very carefully to avoid oxidization, and strain in the metal should be carefully removed by annealing. Even so the contact potential may still vary with temperature and so the temperature should be stabilized. If, for example, in an experiment to measure a gravitational attraction, one of the test masses has an area of the order of 10^3 mm^2, and is separated from the neighbouring surfaces by a distance of 10 mm, and if the temperature coefficient of the potential difference is 10^{-2} V deg^{-1}, the force on the test mass, treating the mass and electrode as a parallel plate capacitor, would change by 4×10^{-15} N deg^{-1}. If the test mass weighed 40 g, the temperature would have to be controlled to 10^{-3} deg to permit accelerations to be measured to 10^{-16} m s^{-2}.

Temperature dependence of the stress in suspension fibre

The equilibrium position of the framework of a torsion balance is determined by the mechanical state of the suspension fibre, which in turn depends upon the internal structure of the fibre. The internal structure of a metal is in general not stable and there are two principal effects. One is creep, the long-time drift in one direction caused by slow relief of stress; the other is thermal stress of the fibre, which makes the equilibrium position of the frame vary with temperature. Both the effects occur because the different parts of a fibre are restrained internally by strain built up when the fibre was made. There will then be a stress which changes with temperature even when there are no external forces.

Because the suspension fibre is under the stress of the load of the frame and test masses, thermal changes produce additional stresses which lead to changes in the structure of the fibre and hence in its equilibrium angular position.

Three conclusions follow from those general considerations:

(1) The thermal stress will be proportional to the coefficient of the thermal expansion of the material; for example, the linear thermal coefficient for tungsten is 4.5×10^{-6} and for quartz is 5.5×10^{-7}. This indicates that the effect caused by thermal stress of a tungsten suspension fibre is higher than that of quartz when the residual stresses are the same.

(2) Because the residual stress in a suspension fibre may be quite complex, it may vary with the cross-section of the wire, it may depend on the load and it may change with time. Consequently the mechanical effect of thermal stress may be highly variable and it is not easily distinguished from random noise. A suspension fibre with high residual stress will cause serious disturbance in the experiment and it is essential to reduce residual stress as far as possible.

(3) The thermal effect only occurs because of the existence of internal residual stress. The equilibrium position of the torsion balance with a stress-free suspension fibre would not change with temperature.

A most important precaution in a gravitational experiment with a torsion balance is to remove the residual stress in the suspension fibre by an annealing procedure. Chen *et al.* (1984) have observed that the pure heat treatment described by Braginsky & Manukin (1977) may not be quite sufficient to remove the residual stress of a tungsten wire, since the melting point of tungsten is very high (about 3200 °C), while the highest annealing temperature that can practically be attained is only just above 1000 °C.

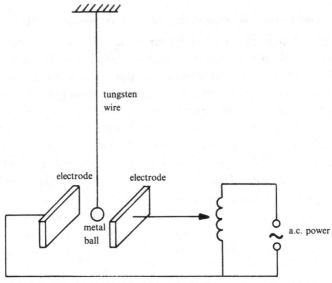

Fig. 3.8 Electrical circuit to drive oscillations of a tungsten wire.

However, Chen (1982) has suggested that a combination of heat treatment and vibration treatment may be used to prepare tungsten fibres. He has used a scheme with four steps.

(1) The wire is hung for a certain length of time (for example one or two weeks) under a load that stresses it very close to its breaking point.

(2) The wire is loaded to a stress of about one-third of the breaking stress and is then subjected to a mechanical vibration at a few tens of hertz. There are many ways of doing that, one of which is shown in Fig. 3.8. A potential difference of about 10 V is maintained between two plates, one on either side of a metal mass hung on the end of a fibre, the gaps between the mass on the end of the fibre and each plate are about 0.1–0.3 mm. When the mass swings towards a plate, before it touches the plate an electrical spark drives it away and it then swings towards the other plate and the process repeats. Thus, the metal mass vibrates forwards and backwards like the clapper of an electrical bell. The voltage across the plates should be very carefully adjusted to ensure that the oscillatory motion of the metal mass does not become too violent, otherwise the fibre will break. The whole mechanism should be housed in a vacuum chamber, in which the air is at a pressure of less then 10^{-2} Pa, for otherwise the tungsten fibre would be oxidized at the temperature to which it is raised by the current passing through it.

Table 3.2. *Procedure for annealing tungsten wire*

Time (min.)	Colour	Temperature (°C)
5	Dark red	600
5	Red	800
10	Red orange	1000
120	Orange	1200
15	Yellow-white	1400
10	Red	800
300	Dark red	600

(3) The third step is to anneal the wire in a vacuum chamber by passing a current through it under same tension as in step (2). The temperature of the wire can be gauged approximately by its colour. Table 3.2 lists such a procedure, the temperature of the wire is raised by steps and at each step is allowed to anneal for the time given in the first column of the table.

(4) After the annealing procedure is completed, the vibration treatment described in (2) is repeated for a period of about 12 h, or more if circumstances allow. With these four steps a very high quality wire may be obtained.

3.6 Air currents

Air currents caused by a temperature gradient inside the housing of a torsion balance have a very important effect on the motion of the torsion frame through horizontal temperature gradients. The force acting on a test mass of surface area A is

$$F = A\Delta p, \tag{3.33}$$

where Δp is the pressure difference caused by the temperature gradient ΔT in question. Δp may be calculated from the ideal gas law so that

$$F = A\frac{R\Delta T}{v}, \tag{3.34}$$

where R is the gas constant and v is the volume of ideal gas at p atm. The volume of 1 kmole of ideal gas at 273 K, 1 atm, is $v_0 = 22.4136$ m^3 kmole^{-1}. Hence the volume of 1 kmole at p atm is $v_0 p_0/p$, where p_0 is 1 atm and eqn (3.34) becomes

$$F = A\frac{Rp}{v_0 p_0}\Delta T. \tag{3.35}$$

If A is 10^4 mm^2, p/p_0 is 1.3×10^{-7} and if ΔT is 10^{-6} deg, the force is about 5×10^{-18} N. If the test mass is 30 g, the disturbing acceleration may be 1.6×10^{-15} m s^{-2}, which is low enough to permit precise gravitational measurements. However, with the same change of temperature ΔT but at atmospheric pressure, the level of disturbance will rise to 1.2×10^{-8} m s^{-2}, a level that is not acceptable for precision measurements. Here we can see the importance of low pressure in gravitational experiments using a torsion balance. None the less, in an early experiment to test the weak principle of equivalence, Eötvös (1891, see section 4.5) did not employ a vacuum but instead used air at 1 atm pressure in a close-fitting jacket. To eliminate temperature gradients, the jacket had three layers of metal wall with air spaces between them. The actual temperature gradient was not recorded but, according to the above discussion, it must have been less than 10^{-9} deg for the precision claimed.

3.7 Local extraneous gravitational attraction on a torsion balance

The effect of local extraneous gravitational attraction on a linear oscillator depends entirely upon the structure of the oscillator. As before, the discussion in this section applies to a torsion balance.

The effect on a torsion balance of moving local bodies has very often been noticed in delicate gravitational experiments. Nearly everyone who has made gravitational experiments has reported such an effect. Although they cannot be reduced by gravitational shielding, there are some measures which can be taken to reduce the gravitational disturbance caused by a local moving body. The first measure is simply to locate the apparatus far away from the moving bodies, for example to install the torsion balance underground (which has other advantages also). Moving bodies affect the balance most seriously when they are in the same horizontal plane, and most movements of local bodies occur at ground level, but the effects cannot be completely eliminated underground and in particular the attraction of the body of an observer, and the gravitational effect of the ground motion (see section 3.2) remain. The second measure takes advantage of the narrow bandwidth of the linear oscillator to limit the extraneous gravitational effects. Disturbances of gravitational sources usually lie in two extreme bandwidths, one very low and the other very high. The gravitational disturbances caused by human activities usually fall in a range from a few hertz to a few tenths of hertz; whereas those caused by environmental changes, such as snowfall outside the laboratory, geometrical changes of nearby masses due to temperature, trees growing

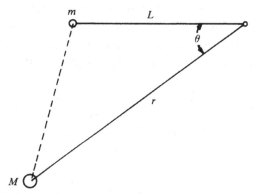

Fig. 3.9 Calculation of the torque exerted by a local mass M on a 'half-arm' torsion balance.

near a laboratory, and so on, lie in a range below 10^{-4} Hz. Hence, if the linear oscillator has a very narrow bandwidth, for example if a resonance method with a modulated gravitational source is used, the extraneous gravitational pulls can be effectively 'filtered'. The third measure is to use the multifold symmetric torsion frame. This framework gives the test masses some high degree of symmetry. We now consider the torque exerted by a concentrated local mass upon a torsion balance possessing some degree of rotational symmetry from which it is possible to understand how the local extraneous gravitational attraction noise can be reduced by using a symmetric torsion balance.

First consider a single half-arm torsion balance, as shown in Fig. 3.9, chosen only for the convenience of calculation. Let the mass at the end of it be m. Let the torque caused by the attraction of a local mass M be τ, let the distance from M to the rotation axis of the torsion balance be r (it is assumed $r > L$), and let the angle between the half-beam L and the line r be θ. In the following derivation it is assumed that the local mass M is located in the same horizontal plane as the torsion framework but if it is not so the necessary correction is straightforward. In any case, the greatest effect is obtained when M is in the horizontal plane of the frame. If the mass M may be assumed to be a point mass the torque acting on the single-beam torsion balance is

$$\tau(\theta) = \frac{GMmL}{r^2}\sin\theta$$

$$\times\left[1 + 3\frac{L}{r}\cos\theta - \frac{3L^2}{2r^2}(1 - 5\cos^2\theta) - \frac{15L^3}{2r^3}(\cos\theta - 7\cos^3\theta) + ...\right], \quad (3.36)$$

It is assumed that the mass of the beam is negligible. The first term represents the gravitational dipole torque, the second term represents the gravitational quadrupole torque, the third term represents the gravitational sextupole torque, the fourth term represents the gravitational octupole, and so on.

This result can easily be extended by superposition to a torsion framework with any particular rotational symmetry, which is why we begin with a study of a 'single-arm' torsion balance. A few results are now listed.

(1) *Single-beam torsion balance with two equal masses at equal distances from the point of support.* The torque exerted on the balance by a local point mass M is

$$\tau_2 = \tau(\theta) + \tau(\theta + \pi), \tag{3.37}$$

and the result is

$$\tau_2 = \frac{GMmL}{r^2} \sin\theta \left[6\frac{L}{r}\cos\theta - 15\frac{L^3}{r^3}(\cos\theta - 7\cos^3\theta) + \ldots \right]. \tag{3.38}$$

The dipole term has disappeared and only the quadrupole, sextupole, octupole and higher multi-pole terms remain.

(2) *Threefold symmetric framework.* The torque in this case may be written as:

$$\tau_3 = \tau(\theta) + \tau(\theta + \tfrac{2}{3}\pi) + \tau(\theta - \tfrac{2}{3}\pi), \tag{3.39}$$

and the result is

$$\tau_3 = \frac{45}{8}\frac{GMmL^3}{r^4}\sin\theta(3\cos^2\theta - \sin^2\theta) + \ldots. \tag{3.40}$$

Only sextupole and higher-order torques remain for a triangular framework; this kind of structure can greatly reduce the noise caused by local moving bodies.

(3) *Fourfold symmetric framework.* Following the same procedure as before, for a fourfold symmetric framework we have

$$\tau_4 = \tau(\theta) + \tau(\theta + \tfrac{1}{2}\pi) + \tau(\theta + \pi) + \tau(\tfrac{3}{2}\pi + \theta); \tag{3.41}$$

the gravitational torque is

$$\tau_4 = \frac{105}{4}\frac{GMmL^4}{r^5}\sin 4\theta + \ldots, \tag{3.42}$$

with only octupole and higher-order torques.

The more symmetric the framework of a torsion balance, the less any extraneous gravitational influence will be, but that may be at the expense of sensitivity to the effect it is desired to detect or measure. Examples of using the multifold symmetric torsion balance are given in section 4.8.

In Chapters 2 and 3, seven types of disturbance have been discussed: thermal noise, ground noise, magnetic disturbance, temperature effects, electrostatic disturbance, air current and local gravitational effect. They do not exhaust all possible disturbances. Other sources of disturbance such as amplifier noise, shot noise and the fluctuation of the light source of an optical lever, the fluctuation of the vacuum and fluctuation of voltage power supply can be present in an experiment and may be quite serious in some special circumstance or in a poorly designed experiment. However, the seven sources discussed in the above are the most common ones. In the following parts of the book, when a specific experiment is discussed, we will use the formulae established here to estimate their effects.

4

The weak principle of equivalence

4.1 Introduction

The essence of the principle of equivalence goes back to Galileo and Newton who asserted that the weight of a body, the force acting on it in a gravitational field, was proportional to its mass, the quantity of matter in it, irrespective of its constitution. This is usually known as the weak principle of equivalence and is the cornerstone of Newtonian gravitational theory and the necessary condition for many other theories of gravitation including the theory of general relativity. In recent times, however, it was found that the weak principle of equivalence was not sufficient to support all theories and the principle has been extended as (1) Einstein's principle of equivalence and (2) the strong principle of equivalence.

Following a brief discussion of the principle of equivalence, this chapter is devoted to an account of the principal experimental studies of the weak principle of equivalence.

4.2 Einstein's principle of equivalence

Gravitation is one of the three fundamental interactions in nature and a question at the heart of the understanding of gravitation is whether or how other fundamental physical forces change in the presence of a gravitational force.

Einstein answered this fundamental question with the assertion that in a non-spinning laboratory falling freely in a gravitational field, the non-gravitational laws of physics do not change. That means that the other two fundamental interactions of physics – the electro-weak force and the strong force between nucleons – all couple in the same way with a gravitational interaction, namely: in a freely falling laboratory, the non-gravitational laws of physics are Lorentz invariant as in the theory of special relativity.

The Einstein principle of equivalence comprises three sub-principles:

(1) The principle of weak equivalence: in a non-spinning freely falling laboratory, the trajectory of a test mass is independent of its internal structure and constitution.

(2) The principle of local Lorentz invariance: in that laboratory, the results of any non-gravitational experiment are independent of the speed of the falling frame.

(3) The principle of local position invariance: in that laboratory, the experimental results of any non-gravitational experiment are independent of the position in space and time of the falling frame.

At first the three principles are independent but Schiff (1970) conjectured that for any complete gravitational theory, some intrinsic connection might exist among them. However, no convincing proof of the conjecture has so far been published.

Einstein's principle of equivalence is most important in the theory of gravitation, particularly in the theory of general relativity. Before we go into the detailed experimental work, it would be appropriate to discuss why that is so.

Any field theory describing the interaction of a field source and a test particle will have two basic equations. One is the field equation to describe the field and how it changes. The other is the equation of motion of a test particle in a given field describing the possible trajectory of the particle. The equation of motion usually has the following form:

$$M_i\left(\frac{\mathrm{d}^2 x^\mu}{\mathrm{d}s^2} + \Gamma^\mu_{\nu\rho}\frac{\mathrm{d}x^\nu}{\mathrm{d}s}\frac{\mathrm{d}x^\rho}{\mathrm{d}s}\right) = AB^\mu, \tag{4.1}$$

where M_i is the inertial mass of the particle, A is a coupling coefficient, B^μ is the field variable, x^μ, x^ν, x^ρ ($\mu, \nu, \rho = 0, 1, 2, 3$) are the coordinates of the particle, $\mathrm{d}s$ is the elementary four-dimensional interval and $\Gamma^\mu_{\nu\rho}$ is the connection coefficient (or the Christoffel symbol of the second kind). For example, in the field of electromagnetism, the equation of motion of a charged particle is

$$M_i\left(\frac{\mathrm{d}^2 x^\mu}{\mathrm{d}s^2} + \Gamma^\mu_{\nu\rho}\frac{\mathrm{d}x^\nu}{\mathrm{d}s}\frac{\mathrm{d}x^\rho}{\mathrm{d}s}\right) = qF^\mu_\nu\frac{\mathrm{d}x^\nu}{\mathrm{d}s}, \tag{4.2}$$

where q is the charge of the particle and F^μ_ν is the field tensor of electromagnetism. If the space is flat, eqn (4.2) reduces to:

$$M_i \frac{\mathrm{d}U^\mu}{\mathrm{d}\tau} = qF_\nu^\mu U^\nu, \qquad (4.3)$$

where U^μ is the four-velocity. In an electromagnetic field, the acceleration of a charged particle depends on its inertial mass and electric charge. On the other hand, according to the Einstein principle of equivalence, all particles in a gravitational field should obey the same laws of physics regardless of their constitution, velocity and position and so the acceleration of a particle should be independent of its inertial mass M_i and the coupling coefficient A in eqn (4.1); that is to say, the ratio of the inertial mass to the coupling coefficient should be a constant. Therefore AB^μ/M_i is a quantity which is solely a function of the field variable. Hence the equation of motion of a particle in a gravitational field may be written as

$$\frac{\mathrm{d}^2 x^\mu}{\mathrm{d}s^2} + \Gamma_{\nu\rho}^\mu \frac{\mathrm{d}x^\nu}{\mathrm{d}s} \frac{\mathrm{d}x^\rho}{\mathrm{d}s} = 0, \qquad (4.4)$$

where $\Gamma_{\nu\rho}^\mu$ is a newly defined connection coefficient which is only a function of the field (different from that in eqn (4.1)). The field variable in the equation of motion has been 'absorbed' into the space connection coefficient, making gravity a purely geometric phenomenon; in this picture, the trajectory of a particle in a gravitational field is a geodesic line in a curved space.

It is clear from the above discussion that the Einstein principle of equivalence is not in any sense derived from other more fundamental principles of physics, neither can it be explained by other theories. It depends on direct experimental tests or on the credibility of the conclusions derived from the theory established on it.

4.3 Strong principle of equivalence

In the previous section dealing with the Einstein principle of equivalence in a freely falling laboratory, only non-gravitational experiments were considered. Is there also a principle of equivalence for gravitational experiments? Can the principle of equivalence be extended to all interactions in nature including gravitation? The so-called strong principle of equivalence is just that extension.

It seems hard to imagine that the following statement could be true: in a non-spinning freely falling laboratory, the outcome of gravitational experiments does not depend upon the position, velocity or constitution of a test body, particularly when the selfmass of the test body is so large or so concentrated that it can easily change the local metric of space-time. The strong principle of equivalence none the less asserts that statement to be true.

As with the Einstein principle of equivalence, the strong principle of equivalence consists of three sub-principles:

(1) *The principle of gravitational weak equivalence.* In experiments involving a test body of large mass, the weak principle of equivalence is still valid. This may be called the gravitational weak equivalence principle. Sufficiently large masses cannot be used in the laboratory and the only hope of performing such experiments is to use astronomical bodies. One of the purposes of lunar laser ranging experiments has been to test the gravitational weak equivalence principle. According to Nordtvedt (1968), if the above principle is not valid, and the ratios of inertial mass to gravitational mass are different for the Earth and the Moon then there will a periodic variation in the distance between the Earth and the Moon: the Nordtvedt effect. Observations are still being made but analysis of the preliminary data has already shown that the gravitational weak equivalence principle is valid to a precision of 7×10^{-12} (Williams *et al.*, 1976).

(2) *The principle of gravitational local Lorentz invariance.* In a non-spinning, freely falling laboratory, the results of any experiment including gravitational experiments are independent of the speed of the frame. The name given to the breaking of this principle is the preferred frame effect. Experiments to search for the preferred frame effect are analogous to ether experiments performed at the end of the last century and may be called gravitational ether experiments.

(3) *The principle of gravitational local position invariance.* In a non-spinning, freely falling laboratory, the results of any experiment including gravitational experiments are independent of the position (both in space and time) of the falling frame. The name given to the breaking of this principle is the preferred position effect. One famous example of the preferred position effect is the speculation concerning the change with time of the gravitational constant. So far this experiment has only been carried out using astronomical observations. But whether it is possible to design an experiment in the laboratory to detect the preferred position effect is still an open question.

It is well known that many new gravitational theories have been proposed in recent decades, some of which, unlike general relativity, predict the violation of the strong principle of equivalence. Tests of the strong principle of equivalence may therefore be very important in testing theories of gravitation.

4.4 Weak principle of equivalence

The weak principle of equivalence is a necessary condition for both the Einstein and the strong principle of equivalence. Besides the statement given in section 4.1, the following are equivalent: 'In a uniform gravitational field, test bodies with different structures and different constitutions have the same acceleration' and 'Test bodies with different structure and different constitutions have the same ratio of inertial mass to gravitational mass'. Also, according to Bondi (1957) and Dicke (1964), the principle may be stated as: 'The ratio of the passive gravitational mass and the inertial mass of a test body is constant.'

Newton (1687) was the first to notice the importance of the equivalence of mass and weight. In the *Principia*, he stated that 'by the help of a pendulum I tried experiments with gold, silver, lead, glass, and common salt, wood, water, and wheat.... By these experiments, in bodies of the same weight, I could manifestly have discovered a difference of matter less than the thousandth part of the whole, had any such been.'

Mass and weight are different aspects of matter. Bondi (1957) carefully distinguished within Newton's concept of mass and weight, three different kinds of mass: inertial mass, passive gravitational mass and active gravitational mass. Inertial mass (M_i) is the mass which appears in Newton's second law, and is inversely proportional to the acceleration of the test body under a constant force. Passive gravitational mass (M_{pg}) is the quantity proportional to the acceleration of the test body under gravitational attraction. Active gravitational mass (M_{ag}) is the source of gravitational attraction.

It is a consequence of Newton's third law (action and reaction are equal and opposite) that passive and active gravitational masses are equal. In modern gravitational theory, however, the equivalence is still an open question. In trying to generalize Gauss's theorem in Newtonian mechanics to the theory of general relativity, Whittaker (1935) found that the natural generalization of gravitational mass in classical theory is the energy tensor in the general theory of relativity. However, the energy tensor does not necessarily contain 'matter' in the classical sense, so that, strictly speaking, there is still a question in modern gravitational theories concerning the validity of the equivalence of these two kinds of mass. Although Kreuzer's experiment (1968) has shown that the ratio of passive to active gravitational mass for fluorine and bromine are the same to within 5 parts in 10^5, further experimental evidence is needed.

To conclude the above discussion, the answer to the question whether the equivalence

$$M_i = M_{pg} = M_{ag} \qquad (4.5)$$

holds or not is an experimental problem.

In the following sections we discuss only experiments on the weak principle of equivalence, i.e. only the experiments concerning the first part of the above equation. Because we are not concerned with the difference between M_{pg} and M_{ag}, in the following, M_{pg} is just simplified as M_g. Since it is obviously not practicable to examine the equivalence of inertial and gravitational mass for every conceivable pair of materials, the question arises whether a result for one particular pair of materials is valid quite generally. Specifically, if the weak principle of equivalence has been verified for a certain pair of materials with high precision, is it possible that there exists some special pair which violates the principle?

4.5 The Eötvös coefficient

Results of tests of the weak principle of equivalence are often expressed in terms of the Eötvös coefficient, the definition and significance of which we now consider.

According to the theory of special relativity, mass M and energy E are equivalent:

$$M = E/c^2, \qquad (4.6)$$

where c is the speed of light. Generally speaking, energy is associated with each of the three fundamental interactions, and for every kind of test mass of whatever material, the total mass is composed of masses created by those interactions in addition to the rest mass. The individual contributions to the mass from the different interactions can be calculated from the structure of the material and so the validity of the weak principle of equivalence can be assessed in relation to the mass contribution by each of the interactions. Consequently the pair of materials on which experiments are done should be chosen to have very different internal structures. The greater the difference of structures, the more general will be the result.

The Eötvös coefficient is defined as

$$\eta = \frac{2\left(\dfrac{M_{ag}}{M_{ai}} - \dfrac{M_{bg}}{M_{bi}}\right)}{\dfrac{M_{ag}}{M_{ai}} + \dfrac{M_{bg}}{M_{bi}}}, \tag{4.7}$$

where M_{ai}, M_{bi}, M_{apg} and M_{bpg} are the inertial masses and passive gravitational masses of the test bodies a and b.

Let a certain interaction x, be assumed to give different contributions to the inertial mass and passive gravitational mass of the test bodies a and b, so that:

$$M_{ag} = M_{ai} + \sum_x \eta_x \frac{E_{ax}}{c^2} \tag{4.8}$$

and

$$M_{bg} = M_{bi} + \sum_x \eta_x \frac{E_{bx}}{c^2}, \tag{4.9}$$

where E_{ax} and E_{bx} are the internal energies of the bodies a and b corresponding to the interaction x and η_x is the Eötvös coefficient for the interaction x. Inserting eqns (4.8) and (4.9) into (4.7), the result will be approximately:

$$\eta = \sum_x \eta_x \left(\frac{E_{ax}}{M_{ai} c^2} - \frac{E_{bx}}{M_{bi} c^2}\right). \tag{4.10}$$

That formula relates the experimental result for bodies a and b to the energies of the different interactions.

Most experiments to test the weak principle of equivalence have been performed on more or less ordinary materials using the simple pendulum (including the rigid or compound pendulum), the torsion balance (including the torsion pendulum), bodies in free fall and the chemical balance. In the following sections we discuss the methods in detail. The weak principle of equivalence has also been tested for free atoms, free neutrons, photons and electrons as described briefly in Chapter 5.

4.6 Direct observation of free fall

The most direct way of testing the weak principle of equivalence is to compare the gravitational acceleration of different materials in a free fall experiment, as Galileo is supposed to have done from the Leaning Tower of Pisa. In principle, any gravimeter measuring g by dropping an object can

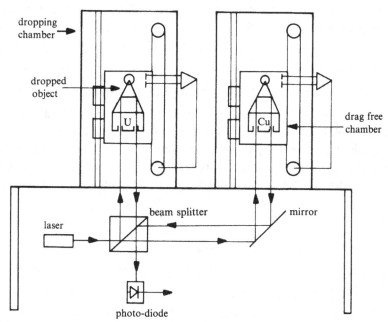

Fig. 4.1 Free-fall method of Niebauer *et al.* (1987) to look for a fifth force.

be used to test the weak principle of equivalence but they may not be able to achieve the precision of a specially designed experiment. An experiment performed by Niebauer, McHugh & Faller (1987) reached a high degree of precision although not as great as that attainable with a torsion balance. The original aim of this experiment was to search for the fifth force (Fischbach *et al.*, 1986) but the result is also a test of the weak principle of equivalence.

As shown in Fig. 4.1, two objects of different materials, copper (weight 40 g) and depleted uranium (weight 199.25 g) were housed in two chambers, evacuated to 10^{-7} torr, and were allowed to fall freely. The difference between the accelerations of the objects of copper and uranium was measured by an interferometer illuminated by a very stable laser.

The incident light was divided in a beam splitter and the two beams were reflected one from each cube-corner reflector fixed to the falling objects. Movements of the interference fringes formed in the recombined beam were detected photoelectrically and the rate of change of phase of one beam relative to the other was measured with an electronic counter. The chambers that contained the two freely falling objects were constrained to move with them; in that way the effect of the resistance of any residual air in the chambers was negligible (see Torge, 1989).

It was found that the difference of acceleration was 0.13 ± 0.50 μGal corresponding to

$$\eta(\mathrm{Cu}, \mathrm{U}) < 5 \times 10^{-10}. \qquad (4.11)$$

4.7 Experiments with the simple pendulum

The history of absolute determinations of gravity with pendulums (see Torge, 1989) shows that precisions of much better than 1 in 10^7 are unobtainable. For that reason, pendulum tests of the principle of equivalence are now of only historical interest, but it is still instructive to consider the possible errors to which such an elementary device is subject, if only as a warning when using more sensitive ones.

Principle

The equation of motion of a simple pendulum of length l in a gravitational field g is, for small amplitudes,

$$m_i l \ddot{\theta} + m_g g \theta = 0, \qquad (4.12)$$

where θ is the angle of deviation of the pendulum, m_i the inertial mass and m_g the gravitational mass. The period is then

$$T = 2\pi \left(\frac{m_i}{m_g}\right)^{\frac{1}{2}} \left(\frac{l}{g}\right)^{\frac{1}{2}}. \qquad (4.13)$$

If two pendulums have the same length l but are made of different materials, then the difference of the squares of the periods is

$$T_1^2 - T_2^2 = \tfrac{1}{2}(T_1^2 + T_2^2)\eta(1, 2), \qquad (4.14)$$

where $\eta(1, 2)$ is the Eötvös coefficient defined in eqn (4.7). Thus

$$\frac{\Delta T^2}{T^2} = \eta(1, 2). \qquad (4.15)$$

Sensitivity of a simple pendulum

In any experiment, the final precision is determined by two factors: the noise level of the experiments (including the environment) and the sensitivity (or the resolution) of the instrument. In this book, we have defined the sensitivity of the device as the smallest acceleration the device can detect if the noise of the system is neglected. Generally speaking, it does not determine the real precision of the apparatus, as the noise level of the linear oscillator may set a higher limit to the smallest measurable acceleration. At the same time, the sensitivity should be matched to the noise level. The acceleration of a simple pendulum of period T and length l swinging with an amplitude x_0 is

$$a = -4\pi^2 l \frac{x_0}{T^2}. \tag{4.16}$$

If the length is kept constant, the change of period corresponding to a change of vertical acceleration Δa is

$$\Delta a = -2gx_0 \frac{\Delta T}{T}. \tag{4.17}$$

So far as the uncertainty of time measurements is concerned $\Delta T/T$ may be as small as 10^{-10}, while x may be as small as 10^{-6}–10^{-4} rad, so that apparently the sensitivity can reach up to 10^{-13} cm s^{-2}. However, the true uncertainty in the period is determined by the precision of locating the instant at which the pendulum passes through its rest position, so that

$$\frac{\Delta T}{T} = \frac{\Delta x}{x_0 n}, \tag{4.18}$$

where Δx is the uncertainty in the angular position of the pendulum and n is the number of oscillations over which the period is measured. Therefore eqn (4.17) is replaced by

$$\Delta a = \frac{8\pi^2 l}{T^2} \frac{\Delta x}{n} = 2g \frac{\Delta x}{n}. \tag{4.19}$$

The uncertainty of the best current measurements of angle is about 10^{-10} rad and if the observation lasts for 100 oscillations, the resolution of a simple pendulum would be 10^{-9} cm s^{-2}.

Possible sources of systematic errors – some aspects of a real simple pendulum

The general discussions of noise on a linear oscillator (Chapters 2 and 3) apply to the simple pendulum; and no more will be said here except to comment on the effect of ground noise on a simple pendulum in tests of the weak principle of equivalence. The horizontal translational motion of the support of a simple pendulum is very serious, but can be greatly reduced by

swinging two pendulums in phase on the same support as was done by
Newton (1687). Since the experimental tests of the weak principle of
equivalence depend on comparing two simple pendulums of different
materials, it is evident that they should be hung from the same support and
set to swing in phase. The difference of periods is then insensitive to the
common motion of the support.

Real pendulums differ from simple pendulums in a number of ways,
some of which we now consider.

Non-linearity

The equation of motion of a pendulum swinging with a finite amplitude is

$$l\ddot{x} + g \sin x = 0, \tag{4.20}$$

which has the solution

$$T = 4\left(\frac{l}{g}\right)^{\frac{1}{2}} K(k), \tag{4.21}$$

where $K(k)$ is the elliptical integral of the first kind, and the parameter k is
$\sin\frac{1}{2}x_0$.

The period of the simple pendulum can thus be written as

$$T = 2\pi\left(\frac{l}{g}\right)^{\frac{1}{2}} (1 + \tfrac{1}{4}\sin^2\tfrac{1}{2}x_0 + \tfrac{9}{64}\sin^4\tfrac{1}{2}x_0 + \tfrac{25}{256}\sin^6\tfrac{1}{2}x_0 + \ldots), \tag{4.22}$$

a formula which shows how the period depends on an amplitude. As an
example, if we consider two pendulums with the same length but swinging
with different amplitudes, 4° and 2°, the fractional difference in their
periods will be 2×10^{-4}. The effect is important and the amplitude should
be measured accurately in order to correct for it.

Effect caused by the yielding of the suspension point

Non-linearity is not the only influence on the relation of amplitude to
frequency and there are other effects which lead to a rather complicated
relation. For that reason we emphasized above that the correction for
amplitude alone should be made using the theoretical eqn (4.22) instead of
experimental results.

The suspension point is not in practice absolutely rigid and will deform
and yield under the action of the pull of the swing bob. A general
discussion is difficult since the yielding will depend not only on the
materials of the support but also on the method of clamping the suspension
wire. Even so we may see that the effect depends on the mass of the bob and
the amplitude.

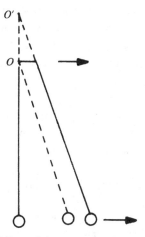

Fig. 4.2 Yielding of the support of a simple pendulum.

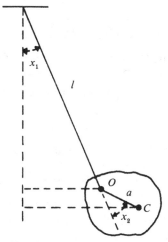

Fig. 4.3 Simple pendulum with bob of finite size.

As shown in Fig. 4.2, a simple pendulum is hung from a point O which is not rigid. Since a varying force is exerted at the point O while the pendulum is swinging, O is not the true centre of rotation. The force acting on the suspension point has vertical and horizontal components. The vertical component will be practically constant if the angle of swing is not large, and so the motion of the suspension point will be essentially horizontal. Thus, from Fig. 4.2, we see that the true suspension point will move to point O', so changing the equivalent length of the pendulum. The change of period depends on the yielding of the support and the amplitude of the swing, but if two pendulums are swung together in phase the difference of the period, which should be proportional to η, will be relatively unaffected by motion of the common support.

Effect caused by the finite size of the bob

Another possible systematic error is caused by the assumption that the bob is a point mass, which is nearly the case. If the bob is of finite size the system becomes a combination of a pure simple pendulum and a compound pendulum.

As shown in Fig. 4.3, we assume that the bob is fixed on the wire at a point O. Although, the attachment of the wire to the bob may be complex, for convenience it is assumed that the point O is an ideal pivot point and there is no frictional loss there. The distance from O to the centre of gravity of the bob C is a, the length of the wire is l and the mass of the wire can be neglected. Under these conditions, the Lagrangian of the whole system will be

$$L = \tfrac{1}{2}ml^2\dot{x}_1^2 + ml a\dot{x}_1\,\dot{x}_2 + \tfrac{1}{2}ma^2\dot{x}_2^2 + \tfrac{1}{2}mR^2\dot{x}_2^2 - \tfrac{1}{2}mglx_1^2 - \tfrac{1}{2}mgax_2^2, \quad (4.23)$$

where m is the mass of the bob and R is the radius of gyration of the bob about the point O.

The equations of motion for x_1 and x_2, from the above Lagrangian, will be

$$l\ddot{x}_1 + a\ddot{x}_2 + gx_1 = 0, \quad (4.24)$$

$$(R^2 + a^2)\ddot{x}_2 + la\ddot{x}_1 + gax_2 = 0. \quad (4.25)$$

The frequency ω of the resulting motions is given by the following formula:

$$\omega^2 = \frac{la + a^2 + R^2}{2R^2}\left\{1 + \left[1 - \frac{4alR^2}{(la + a^2 + R^2)^2}\right]^{\frac{1}{2}}\right\}\frac{g}{l}. \quad (4.26)$$

This equation shows how the size of the bob affects the frequency. Notice that, even with a finite size of bob, the effect can vanish if it is arranged that a is zero, for then eqn (4.26) reduces to

$$\omega^2 = \frac{g}{l}. \quad (4.27)$$

In practice, it is not easy to arrange for a to be zero. Notice also that the frequency of the pendulum depends on R, the radius of gyration of the bob about its pivot. But R is a function of the mass distribution of the bob. This may lead to a difference between pendulums of the same mass but different materials, for example if one is a liquid (Bessel's experiment (1832a), this section). Furthermore changes of temperature alter the length of the pendulum, the position of the centre of gravity (as most authors have already noticed), and the radius of gyration, and so also lead to differences of period between pendulums of different materials.

In summary, the uncertainties in the determination of $\Delta T^2/T^2$ (eqn (4.15)) are of two sorts, the uncertainty of measurement of T and the disturbances due to ground motion, elasticity of the support, the finite amplitude of the pendulum, and so on. The uncertainty of measurement is nowadays much less than that due to the other causes, while they can be reduced by having two pendulums of different materials but similar periods oscillating on the same support (see, for example, Torge (1989) for a discussion of some of the principal effects). The main problem in the use of simple pendulums is to construct two pendulums of the same effective length but of different materials. Eqn (4.26) shows that the length l should be great relative to the size of the bob of the pendulum, as with the pendulums used by Newton (below).

Experiments

Historical records show that (albeit unconsciously), Galileo in 1638 was the first to use a simple pendulum to verify the weak principle of equivalence. He once studied two simple pendulums of the same length, one having a ball of lead, the other a ball of cork. Although the lead ball was 100 times heavier than the cork ball, Galileo found that the pendulums had the same period. He started the pendulums swinging at the same time and found that after more than one hundred oscillations, they were still keeping well in phase. The corresponding Eötvös coefficient for lead and cork would be

$$\eta < 2 \times 10^{-3}. \tag{4.28}$$

Newton was the first consciously to use the simple pendulum to test the weak principle of equivalence. He took particular account of the resistance of the air. Two boxes of the same size were hung by a suspension wire 3.3 m in length from the same support and in them he placed different materials such as gold, silver, lead, glass, salt, wood, water and wheat. His result can be interpreted as

$$\eta < 10^{-3} \tag{4.29}$$

for any of the combinations.

Bessel (1832b, 1889) repeated Newton's experiment using a simple pendulum. The development of more accurate measurements of time and length in the early nineteenth century made Bessel decide to use two simple pendulums with different lengths instead of the same length. The materials of the test bodies he used were iron, zinc, lead, silver, gold, triferric tetroxide, marble, clay, quartz and water. He found that for all pairs of those materials, except those with water, the Eötvös coefficient was less than 10^{-5}. The exceptional behaviour of water is probably to be ascribed to internal motions.

Thomson (1909) also used a simple pendulum to test whether the weak principle of equivalence would be violated between radioactive and non-radioactive matter: he had estimated from an ether theory that the ratio of the inertial mass to the weight of radioactive matter may be greater than that of non-radioactive matter by $1/13\,000$. Southerns (1910) later designed an experiment to check the hypothesis by comparing uranium oxide and lead oxide. He used a simple pendulum and a compound pendulum and his result indicated that the ratio of inertial mass to weight for uranium oxide does not differ from that of lead oxide by more than 5×10^{-6}.

Potter (1923) measured the relative variation of acceleration for different materials using a simple pendulum consisting of an agate knife-edge, a steel suspension wire about 900 mm long, and a hollow cylindrical bob which could hold in turn the different materials to be tested. The period was compared with the period of a standard compound pendulum, similar to the first one, except that the steel suspension wire was replaced by an invar rod to make the pendulum rigid. The bob was a brass cylinder. To eliminate the effect of temperature on the brass, the invar rod was secured to the mid-point of the bob. The change in length of the simple pendulum was monitored using a micrometer underneath the cylinder in the following manner: a fine steel ball was attached to the bottom of the cylindrical bob, this ball being located just above the centre of an optically polished plane at the end of the micrometer; the change in relative position of the ball and its image in the polished surface could be read using a high power microscope to a resolution of about 1 μm. The period of the simple pendulum was measured by comparing it with the period of the compound pendulum. Thus Potter was able to compare the acceleration of different materials carried by the simple pendulum relative to the brass of the compound pendulum. His results and the Eötvös coefficients calculated from them are listed in Table 4.1. No significant meaning attaches to the observation that materials like ammonium fluoride, paraffin wax and mahogany have a slightly greater gravitational acceleration than the

Table 4.1. *Potter's tests of the weak principle of equivalence (1923)*

Substances	Acceleration	η
Brass	$1.000\,000\,g$	—
Lead	$0.999\,992\,g$	8×10^{-6}
Steel	$0.999\,980\,g$	2×10^{-5}
Ammonium fluoride	$1.000\,005\,g$	5×10^{-6}
Paraffin wax	$1.000\,013\,g$	1×10^{-5}
Mahogany	$1.000\,015\,g$	1×10^{-5}
Duralumin	$0.999\,992\,g$	8×10^{-6}

others. Another more accurate experiment with a torsion pendulum showed that there is no difference in the mass–weight ratios for these three substances greater than 1 part in 10^6 (Potter, 1927).

4.8 Principle of torsion balance experiments

In simple pendulum experiments, the acceleration is that of the pendulum rotating about its suspension point, while the gravitational field is that of the Earth, but the resolution of that type of experiment is not great. To compare extremely weak forces, Eötvös (1891) developed a new type of experiment using the very small torsional restoring force of a thin suspension fibre.

The torsion balance is a far more sensitive detector of mechanical forces than the simple pendulum and is not subject to defects of design as is the simple pendulum. On the other hand, the forces to which it is subject act in the horizontal plane, so that it is not sensitive to the direct gravitational attraction of the Earth. The principle of the torsion balance for testing the weak principle of equivalence can be briefly summarized as follows. Two different kinds of test body are attached to the ends of the beams of the balance which may be a single beam or a complex but symmetric framework suspended by a fine fibre. The test bodies are subject not only to the gravitational attractions related to their passive gravitational masses, but also to inertial forces related to their inertial masses. Detailed mathematical analysis shows that if the ratios of the gravitational mass to the inertial mass are not the same for the test bodies, there will be a torque about the direction of the suspension fibre acting on the torsion balance to make it rotate. The result of measuring this torque will be a test of the weak principle of equivalence for the two materials.

There are two ways to employ a torsion balance. In one, the horizontal component of the solar attraction is balanced against the corresponding component of the acceleration of the Earth in its orbit about the Sun, while in the other scheme, the acceleration of the Earth spinning about its axis is balanced against a residual gravitational force.

The solar scheme is the simplest to explain. The gravitational attraction of the Earth as a whole to the Sun exactly balances the acceleration of the centre of mass of the Earth in its orbit about the Sun, but at any point on the surface the gravitational force departs from the orbital acceleration by the tide-raising force. Since the tide-raising force is the difference between inertial and gravitational components, it depends on the ratio of inertial to gravitational mass of any test object at the surface of the Earth, and thus if two objects have different ratios, the tide-raising forces upon them will differ. If the test objects are attached to the beam of a torsion balance there may be a periodic variation as the Earth spins upon its axis.

The terrestrial scheme depends upon the fact that the horizontal plane is tangential to a surface that is an equipotential of the combined potential of the Earth's gravitational attraction and the centrifugal force of the spin of the Earth about its axis. The direction of the vertical (in which the torsion balance hangs) thus makes a small angle with the direction of the gravitational attraction, an angle which will depend upon the ratio of inertial to gravitational mass of a test object. If the ratios differ for test objects on the beam of a torsion balance, there will be a net torque on the balance.

Unfortunately, the net torque in the terrestrial scheme does not vary with the rotation of the Earth, and can only be detected by rotating the whole apparatus around the vertical. The solar torque, on the other hand, shows up as a periodic motion of a balance which remains in a fixed orientation.

Consider an object labelled n at one end of the beam of a torsion balance (n is also the number of beams of the torsion frame), and let its position vector relative to the point of suspension be L_n. Let its inertial mass be M_{ni} and its gravitational mass M_{ng}. Let the gravitational acceleration acting on it be g and the inertial acceleration be a (they may vary with the position of the object). The total torque acting on the torsion balance will be the vector product

$$\tau = \sum M_{ni} \left(\frac{M_{ng}}{M_{ni}} g + a \right) \times L_n. \tag{4.30}$$

In order to calculate this vector product, it is necessary to establish a

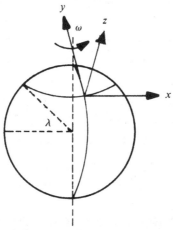

Fig. 4.4 Torsion balance coordinate system (x, y, z) (x: latitude).

convenient coordinate system. Since only the component about the direction of the torsion fibre is of interest in these experiments, we define an appropriate coordinate system attached to the torsion balance, with its z-axis coincident with the direction of the suspension fibre, its x-axis parallel to the latitude circle on the Earth, and its y-axis towards the north pole of the Earth (Fig. 4.4). That system will be used throughout this chapter for calculating the torque. We call it the torsion balance coordinate system. For any design of the torsion beam framework, the task of calculating the torque consists of writing down the vectors g, a and L_n in the torsion balance coordinate system (from the definition of the coordinate system, we know that the L_n always remain in the (x, y)-plane). The component of the vector product in the z-direction (eqn (4.30)) is the torque deflecting the torsion balance: if m is a unit vector in the direction of the suspension fibre of the torsion pendulum (which is the direction of the resultant force – gravitational plus inertial – upon the beam), the torque about the fibre is

$$\sum M_{ni} \left(\frac{M_{ng}}{M_{ni}} g + a \right) \times L_n \cdot m.$$

The total torque exerted on the balance is the sum of such triple products taken over all test objects attached to the balance, and the analysis of the performance of any balance follows from the evaluation of that sum.

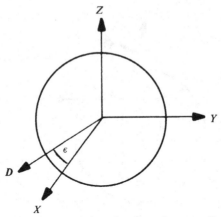

Fig. 4.5 Vector distance D in solar Cartesian coordinate system.

Solar experiments

In this class of experiment, the gravitational attractions of the Sun upon two kinds of test object are compared with the centrifugal forces produced by the orbital motion of the Earth. To obtain an expression for these two attractions, we note that the gravitational force and the centrifugal force act in opposite directions. The gravitational acceleration toward the Sun is

$$g = \frac{GM}{D^3}D, \tag{4.31}$$

where G is the gravitational constant, M is the mass of the Sun, D is the vector from the centre of the Earth to the Sun (or the distance from the torsion balance to the Sun, as the radius of the Earth is negligible compared to this distance). The orbital acceleration a, from Kepler's law, is very nearly

$$a = -\frac{GM}{D^3}D, \tag{4.32}$$

and thus, when the masses of the test objects are all nearly the same, i.e. $M_{1i} = M_{2i} = \ldots = M_{i}$, eqn (4.30) can be reduced to

$$\tau = M_i \Omega^2 \eta D \times \Gamma, \tag{4.33}$$

where

$$\Gamma = \sum \left(\frac{M_{ng}}{M_i} - 1 \right) L_n. \tag{4.34}$$

In order to obtain an explicit expression for the torque, we need the components of D in the torsion balance coordinate system. In Fig. 4.5, let ε be the obliquity of the ecliptic, the angle of intersection between the equator and the ecliptic. In a solar Cartesian coordinate system (X, Y, Z), D is given by

$$D = ID \cos \varepsilon \sin \Omega t + JD \cos \varepsilon \cos \Omega t + KD \sin \varepsilon, \qquad (4.35)$$

where Ω is the angular velocity of the Earth in its orbit, and I, J and K are the unit vectors along the X, Y and Z axes of the solar coordinate system. To find the components of D in the torsion balance coordinate system, we require the transformation matrix between the solar Cartesian coordinate (X, Y, Z) and torsion balance coordinate (x, y, z). There are four differences between the X, Y, Z and x, y, z systems:

(1) The z-axis is inclined to the Z-axis at an angle of $(\frac{1}{2}\pi - \lambda)$, where λ is the latitude (Fig. 4.4); the difference between geographical and astronomical latitude may be ignored.
(2) The X, Y, Z system is fixed in the solar system, but the x, y, z system is fixed on the surface of the Earth, so that the latter rotates about the Z-axis with an angular velocity ω, the angular velocity of the Earth about its polar axis.
(3) From the definition of the x, y, z system, the x-axis is parallel to the latitude circle, while the (X, Y)-plane in the solar system is in the equatorial plane. Consequently, the x-axis remains in a plane parallel to the (X, Y)-plane of the solar Cartesian coordinate system.
(4) The origins of the two systems are displaced by a vector R, the position vector of the balance in relation to the centre of mass of the Earth, but that may be ignored by comparison with the distance to the Sun.

We see that the coordinate transformation is the result of two rotations, the first about the X-axis through an angle $(\frac{1}{2}\pi - \lambda)$, and the second about the Z-axis through an angle ωt. The matrix of the transformation between the two coordinate systems is

$$\begin{pmatrix} \cos \omega t & \sin \omega t & 0 \\ -\sin \lambda \sin \omega t & \sin \lambda \cos \omega t & \cos \lambda \\ \cos \lambda \sin \omega t & -\cos \lambda \cos \omega t & \sin \lambda \end{pmatrix},$$

and it follows that the vector D in the torsion balance coordinate system is

$$D = (D \cos \varepsilon \sin \Omega t \cos \omega t + D \cos \varepsilon \cos \Omega t \sin \omega t) i$$
$$+ (-D \cos \varepsilon \sin \Omega t \sin \lambda \sin \omega t + D \cos \varepsilon \cos \Omega t \sin \lambda \cos \omega t + D \sin \varepsilon \cos \lambda) j$$
$$+ (D \cos \varepsilon \sin \Omega t \cos \lambda \sin \omega t - D \cos \varepsilon \cos \Omega t \cos \lambda \cos \omega t + D \sin \varepsilon \sin \lambda) k,$$
$$(4.36)$$

where i, j, k are the unit vectors along the x, y and z axes of the torsion balance coordinate system. It is now possible to calculate the torque acting on a torsion balance of any configuration. Three configurations have been of particular importance.

(1) *Single beam with two test masses.* In this case, because the two halves of the beam point in opposite directions, eqn (4.34) gives

$$\boldsymbol{\Gamma} = \eta \boldsymbol{L} = \eta L(\cos\phi \boldsymbol{i} + \sin\phi \boldsymbol{j}), \qquad (4.37)$$

from which it follows that the torque (see eqn (4.33)) is

$$\boldsymbol{\tau} = M_1 \Omega^2 \eta \boldsymbol{D} \times \boldsymbol{L}. \qquad (4.38)$$

Combining eqns (4.36), (4.37) and (4.38), the component of the torque in the k-direction will be

$$\tau = M_1 \Omega^2 \eta L D \cos\varepsilon [\sin(\Omega+\omega)t \sin\phi$$
$$+ \sin\lambda\cos(\Omega+\omega)t\cos\phi - \tan\varepsilon\cos\lambda\cos\phi]. \quad (4.39)$$

We see that this torque is a function of the time of year (Ωt), of the time of day (ωt), the latitude (λ), and the orientation (ϕ). Because Ω is much less than ω, the dominant periodic term is about 24 hours.

If ϕ is $\frac{1}{2}\pi$ (the beam is oriented along the north–south direction), eqn (4.39) will reduce to

$$\tau = M_1 \Omega^2 \eta L D \cos\varepsilon \sin(\Omega+\omega)t; \qquad (4.40)$$

the torque will be independent of latitude, and this orientation gives the maximum torque.

If ϕ is 0 (the beam is oriented along the east–west direction), the variable part of the torque will be

$$\tau = M_1 \Omega^2 \eta L D \cos\varepsilon \sin\lambda\cos(\Omega+\omega)t, \qquad (4.41)$$

a function of the latitude. The balance located at the north pole will experience the greatest torque.

A comparison of the maximum torque in a terrestrial experiment (to be discussed later) and that in a solar experiment using the same single-beam torsion balance is important. The ratio of the two torques, from eqns (4.41) and (4.55) is

$$\frac{\omega^2 R \sin\lambda\cos\lambda}{\Omega^2 D \cos\varepsilon}.$$

The numerical value is about 2.9 in latitude 45°. Thus, although the attraction of the Sun is much less than that of the Earth on a test body of

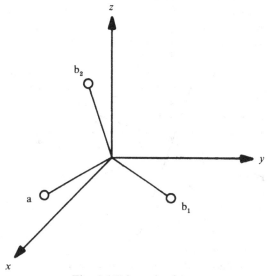

Fig. 4.6 Triangular frame.

a torsion balance, there is not much difference between the expected torques in the terrestrial and the solar experiments, another advantage of using a solar experiment in tests of weak principle of equivalence.

(2) *Threefold symmetric frame work.* As shown in Fig. 4.6, a horizontal triangular frame carrying three test bodies is suspended from its centre by a torsion fibre. A test body of a certain material (a) is placed at one vertex and test bodies of another kind of material (b) are placed at the other two vertices. The beam vectors of the threefold symmetric framework can be written as

$$L_a = L\cos\phi\,i + L\sin\phi\,j, \tag{4.42}$$

$$L_{b_1} = L\cos(\phi + \tfrac{2}{3}\pi)\,i + L\sin(\phi + \tfrac{2}{3}\pi)\,j, \tag{4.43}$$

$$L_{b_2} = L\cos(\phi - \tfrac{2}{3}\pi)\,i + L\sin(\phi - \tfrac{2}{3}\pi)\,j. \tag{4.44}$$

Inserting the above relations into eqn (4.34), it is found that

$$\Gamma = \eta L(\cos\phi\,i + \sin\phi\,j), \tag{4.45}$$

which is exactly the same expression as for the single beam (see eqn (4.37)); no further discussion is needed.

(3) *Eightfold symmetric framework.* This is a star-like assembly of eight test bodies, four of material a on one side of the assembly, and four of

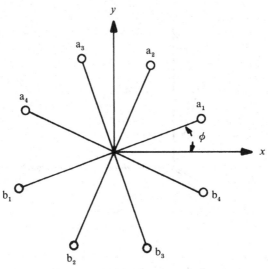

Fig. 4.7 Eightfold symmetric frame.

material b on the other side. As before, to calculate the torque, we must first calculate the vector $\boldsymbol{\Gamma}$ which is determined by the beam vectors. From Fig. 4.7, we may write the vectors in the following form:

$$L_{a_1} = L(\cos\phi\,\boldsymbol{i} + \sin\phi\,\boldsymbol{j}), \tag{4.46a}$$

$$L_{a_2} = L[\cos(\phi + \tfrac{1}{4}\pi)\,\boldsymbol{i} + \sin(\phi + \tfrac{1}{4}\pi)\,\boldsymbol{j}], \tag{4.46b}$$

$$L_{a_3} = L[\cos(\phi + \tfrac{1}{2}\pi)\,\boldsymbol{i} + \sin(\phi + \tfrac{1}{2}\pi)\,\boldsymbol{j}], \tag{4.46c}$$

$$L_{a_4} = L[\cos(\phi + \tfrac{3}{4}\pi)\,\boldsymbol{i} + \sin(\phi + \tfrac{3}{4}\pi)\,\boldsymbol{j}], \tag{4.46d}$$

$$L_{b_1} = L[\cos(\phi + \pi)\,\boldsymbol{i} + \sin(\phi + \pi)\,\boldsymbol{j}], \tag{4.46e}$$

$$L_{b_2} = L[\cos(\phi + \tfrac{5}{4}\pi)\,\boldsymbol{i} + \sin(\phi + \tfrac{5}{4}\pi)\,\boldsymbol{j}], \tag{4.46f}$$

$$L_{b_3} = L[\cos(\phi + \tfrac{3}{2}\pi)\,\boldsymbol{i} + \sin(\phi + \tfrac{3}{2}\pi)\,\boldsymbol{j}], \tag{4.46g}$$

$$L_{b_4} = L[\cos(\phi + \tfrac{7}{4}\pi)\,\boldsymbol{i} + \sin(\phi + \tfrac{7}{4}\pi)\,\boldsymbol{j}]. \tag{4.46h}$$

It is now easy to show that the vector $\boldsymbol{\Gamma}$ is

$$\boldsymbol{\Gamma} = \eta L\{[\cos\phi - (1 + 2^{\frac{1}{2}})\sin\phi]\,\boldsymbol{i} + [\sin\phi + (1 + 2^{\frac{1}{2}})\cos\phi]\,\boldsymbol{j}\}. \tag{4.47}$$

As before, we obtain the magnitude of the torque in the direction of $\boldsymbol{k}$ as

$$\begin{aligned}
\tau = M_i\,\Omega^2\eta LD\cos\varepsilon\{&\sin(\Omega + \omega)\,t\,[\sin\phi + (1 + 2^{\frac{1}{2}})\cos\phi] \\
&+ \sin\lambda\cos(\Omega + \omega)t\,[\cos\phi - (1 + 2^{\frac{1}{2}})\sin\phi] \\
&- \tan\varepsilon\cos\lambda[\cos\phi - (1 + 2^{\frac{1}{2}})\sin\phi]\}.
\end{aligned} \tag{4.48}$$

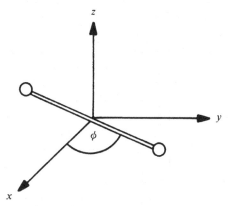

Fig. 4.8 Single-beam torsion balance.

For the special case of $\phi = \frac{1}{8}\pi$, the expression simplifies to

$$\tau = M_i \Omega^2 \eta L D \alpha \cos \varepsilon \sin (\Omega + \omega)t, \qquad (4.49)$$

where α is the factor $[\frac{1}{2}(1-2^{-\frac{1}{2}})]^{\frac{1}{2}} + (1+2^{\frac{1}{2}})[\frac{1}{2}(1+2^{-\frac{1}{2}})]^{\frac{1}{2}}$ (about 2.613). From Fig. 4.7 it is found that, if $\phi = \frac{1}{8}\pi$, the whole assembly is symmetric about the y-axis; this is equivalent to a single beam with a length of $2.613L$, oriented along the north–south direction, a result independent of the latitude of the apparatus just as for a single beam.

We have now discussed three designs. Results for other types of framework can be calculated by the same method and we can prove in general that, for any kind of symmetric framework, placed symmetrically about the north–south direction, the torque acting on the torsion balance is equivalent to that on a single beam, the only difference being a numerical factor (like the number 2.613).

Terrestrial torsion balance experiments

In the terrestrial experiment, g is the gravitational attraction directed to the centre of the Earth, a is the centrifugal acceleration perpendicular to the polar axis of the Earth. The vectors L_n are determined by the type of framework of the torsion balance. In the special case of a single beam, as shown in Fig. 4.8, with two equal test bodies a and b, $M_{ai} = M_{bi} = M_i$ and $L_a = -L_b = L$. The total torque is

$$\tau = M_i \eta g \times L. \qquad (4.50)$$

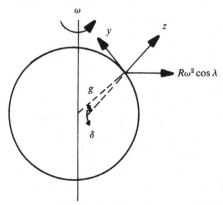

Fig. 4.9 Forces acting on test mass in terrestrial experiment.

It is easy to write down the formulae for g and L in the torsion balance coordinate system as shown in Fig. 4.9. They are:

$$g = -g \sin \delta j + g \cos \delta k, \qquad (4.51)$$

$$L = L \cos \phi i + L \sin \phi j, \qquad (4.52)$$

where δ is the angle between the true gravitational attraction of the Earth and the z-axis, which varies with the latitude of the location of the experimental apparatus, ϕ is the angle of the half-arm of the beam with the x-axis, i, j and k are the unit vectors along the x, y and z axes. The torque in the k-direction is

$$\tau = M_i \eta g L \sin \delta \cos \phi. \qquad (4.53)$$

The angle δ is calculated from the formula

$$\sin \delta = \frac{\omega^2 R \sin \lambda \cos \lambda}{g}, \qquad (4.54)$$

where ω is the angular velocity of the rotation of the Earth about its polar axis, R is the radius of the Earth and λ is the latitude of the location of the torsion balance. Therefore

$$\tau = M_i \eta \omega^2 R L \sin \lambda \cos \lambda \cos \phi, \qquad (4.55)$$

which indicates that if the Eötvös coefficient is not zero, there will be a torque acting on the torsion balance that will vanish if the beam is oriented in the east–west direction ($\phi = 0$ or $\phi = \pi$). However, from an experimental viewpoint, since there is no way in practice of identifying a small constant torque, ϕ must be changed to alter the value of the torque τ. One way of doing this is to interchange the two test bodies, which is equivalent to changing the angle ϕ by 180°. The other is to rotate the whole experimental apparatus to change the angle ϕ. The first method is not practical in most experiments, for the balance must be maintained in a sealed enclosure. Thus in practice the whole balance in its enclosure is turned through 180°, even though with a delicate instrument that is undesirable. That is the principle of the terrestrial torsion balance experiment to test the weak principle of equivalence.

Expected value of the deflection angle

We now summarize the dependence of the deflection of a torsion balance (of period T) upon the Eötvös coefficient of a pair of test materials.

In a terrestrial experiment, with a single-beam balance oriented east–west, the expected value of the deflection angle is, neglecting the mass of the beam,

$$\theta_0 = \frac{\eta \omega^2 T^2 R \sin \lambda \cos \lambda}{4\pi^2 L}. \tag{4.56}$$

We see that the period of the torsion balance is very important for determining the expected value of the deflection: the longer the period, the greater the angle and the higher the precision of the experiment.

In a solar experiment, with a single-beam balance oriented north–south, the expected value of the deflection is, again neglecting the mass of the beam,

$$\theta_0 = \frac{\eta \Omega^2 T^2 D \cos \varepsilon}{8\pi^2 L}. \tag{4.57}$$

If a symmetrical triangular balance is used with the same north–south orientation, the result will be

$$\theta_0 = \frac{\eta \Omega^2 T^2 D \cos \varepsilon}{12\pi^2 L}. \tag{4.58}$$

Similarly, for an eightfold symmetric framework torsion balance, oriented so that $\phi = \frac{1}{8}\pi$, and with the mass of the beam negligible, the expected value of the deflection is

$$\theta_0 = \frac{\eta \Omega^2 T^2 D \alpha \cos \varepsilon}{32\pi^2 L}. \tag{4.59}$$

It follows from eqns (4.57), (4.58) and (4.59) that the more symmetric the framework, the less the expected value of the deflection. However, the more symmetric framework has the merit of reducing the local gravitational influence as already discussed in section 3.7.

The above expressions relating the deflection to the Eötvös coefficient provide the criteria for the design of the detector system and they are the formulae used to calculate the precision of the experiment.

Sensitivity of the torsion balance

The deflection θ of a torsion balance, subject to an acceleration a is given approximately by

$$a = \frac{k\theta}{\sum Lm_i}, \qquad (4.60)$$

where k is the torsion constant, L is the length of the beam and m_i is the mass of the test body attached to the torsion balance. To obtain the expression for the sensitivity of the torsion balance, we may differentiate the above expression with respect to θ, which is the quantity measured in the experiment. Then, we can see that the sensitivity is determined by the torsion constant and the angular resolution of the detection system:

$$\Delta a = \frac{k\Delta\theta}{\sum Lm_i}. \qquad (4.61)$$

In a typical experiment k is 10^{-10} N m rad^{-1}, $\Delta\theta$ is 10^{-10} rad, L is 0.1 m and m is 10 g, and a sensitivity of 10^{-14} cm s^{-2} may be achieved. To reach a better sensitivity the main problems are to decrease the torsion constant and increase the resolution of the angular measurement (for a smaller $\Delta\theta$). It is important to note that an absolute calibration of the angular deflection is not required and in consequence it may in future be possible to obtain significantly better angular resolution perhaps approaching 10^{-12} rad.

4.9 Measurement of torque on a torsion balance

Tests of the weak principle of equivalence demand the most precise measurements possible of the torque on a torsion balance; the possible methods for such measurements are briefly summarized. The torque exerted upon a torsion balance can be measured with the balance set in an open-loop mode or in a closed-loop mode. In the first, the balance is in free motion and the torque is proportional to its average displacement from the

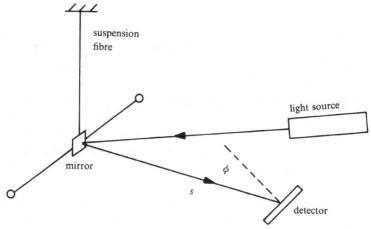

Fig. 4.10 Optical lever to measure angle of deflection of torsion balance.

equilibrium position; whereas in a closed-loop mode the balance is maintained at the equilibrium position and the torque needed to keep it there is found from the response of the control system.

Motion of balance – open loop

The torque exerted on the balance is the product of the torque constant and the angle of rotation. When the torsion balance is set in an open-loop mode, both have to be measured.

(1) *The measurement of the angle of deflection.* Because the torsion balance is a very delicate device, some non-contact method must be used in measuring the motion of the various devices (such as optical levers, electrical transducers and optical interferometers) that may be employed to measure the angle of deflection. The optical lever, shown in Fig. 4.10, is the most commonly used, as it is convenient and reliable. Light is incident on a mirror attached to the middle of the beam or the fibre and is reflected onto a suitable detector at a distance s from the fibre. The deflection angle can be measured by the movement of the spot of the reflected light. It is easy to see that, for a small deflection angle, the power of the amplification of the optical lever is linearly proportional not only to s, but also to the factor $1/\cos\phi$, where ϕ is the angle between the normal to the surface of the receiver and the incident ray.

Although the deflection of the reflected light has been measured with a telescope or photographically, it is more convenient to use some form of

Fig. 4.11 Differential capacitative displacement transducer with two fixed plates.

optoelectric detector, such as the silicon photo cell position sensor. The angular resolution of the optical lever, the smallest deflection of the torsion frame that can be detected, is

$$\Delta\theta = \frac{2d\cos\phi}{s}, \qquad (4.62)$$

where d is the smallest linear displacement that the receiver can distinguish. For example, with a silicon position sensor, d can be as small as 0.1 nm, and so for normal incidence at a distance s of 2 m a resolution of 10^{-10} rad can be reached. Small angles of deflection can also be measured by electrical displacement transducers which usually consist of two capacitors or inductors. Thus, as illustrated in Fig. 4.11, two fixed parallel plates mounted horizontally are placed very close to a counterweight of the torsion balance. When the torsion balance rotates, the changes of C_1 and C_2 in Fig. 4.11 are given by

$$C_1 = C_0 \frac{t_0}{t_0 - \Delta t}, \qquad (4.63)$$

$$C_2 = C_0 \frac{t_0}{t_0 + \Delta t}, \qquad (4.64)$$

where Δt is the linear displacement at the end of the beam of the torsion balance, C_0 is the capacitance when Δt is 0, often less than 10 pF. Various bridge circuits, particularly a transformer ratio bridge (known as TRB), are available for the measurement of very small changes of capacitance (Hague & Foord, 1971; Tohes & Richards, 1973).

Inaccuracies of the capacitative transducer may come from stray fields around the edges of the plates, as has been discussed by several authors (Jones & Richards, 1973).

Optical interferometers may also be used to measure extremely small displacements. However, because of the complexity of the equipment, the fluctuation of the light pressure, and the instability of the torsion balance (due to modes of oscillation other than the rotational mode which cause the optical alignment to be unstable), interferometric techniques are not easy to apply in weak equivalence tests with a torsion balance but they have been used in other gravitational experiments, as will be seen later.

(2) *The measurement of the torsion constant.* The torsion constant is an intrinsic property of a torsion fibre and for an ideal uniform circular fibre of radius r and length l it is

$$k = \frac{\pi r^4 \mu}{2l}, \tag{4.65}$$

where μ is the shear modulus of the fibre. In practice, neither the shear modulus μ nor the radius of the wire (which may vary throughout the length of the fibre) can be known accurately and so k cannot be calculated at all precisely from this formula and must be found from the dynamical properties of the pendulum. The equation of motion of a torsion pendulum in a constant potential is

$$I\ddot{\theta} + k\theta = 0, \tag{4.66}$$

where I is the moment of inertia of the torsion pendulum and θ the deflection angle. The torsion constant is therefore

$$k = \frac{4\pi^2 I}{T^2} = I\omega^2, \tag{4.67}$$

where T and ω are the periodic time and the corresponding angular frequency of the oscillations of the pendulum.

Careful measurement of the period and the moment of inertia of the framework are needed for a precise value of the torsion constant. The method is used not only in tests of the weak principle of equivalence, but also in measurements of the gravitational constant and in other experiments to test the law of gravitation where absolute measurements are necessary. However, it is not easy to measure the moment of inertia for it depends both upon the mass and upon the structure and mass distribution of the torsion system. In the later part of this book we discuss various methods by which it can be measured precisely.

These problems are avoided if the torsion balance is used in a closed-loop mode, for the response to a known torque applied to the balance can be found.

Motion of balance – closed loop

When the balance is controlled by a servo-system, the torque can be measured directly by the external torque exerted on the balance. Electrostatic forces are usually used for that purpose. While there are many

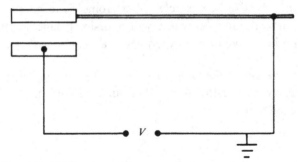

Fig. 4.12 Application of electrostatic force to balance torque.

different ways of applying an electrostatic force to the torsion balance, the principles are the same.

A flat plate placed close to one end of the beam of the torsion balance, as shown in Fig. 4.12, forms a parallel-plate capicator with a flat plate on the balance arm. Let the torsion balance be earthed and a voltage V be applied to the fixed plate so that, in vacuum, the force between them is

$$F = \frac{C^2}{2\varepsilon_0^2 A} V^2, \tag{4.68}$$

where C is the capacitance, A is the area of the plate and ε_0 is the dielectric constant of free space. A force (or torque) proportional to the square of the voltage is not convenient and so a different scheme (shown in Fig. 4.13) that makes the applied force linearly proportional to the voltage is better. Two fixed plates are used on either side of the plate on the balance and are connected to the two poles of a battery of voltage V_0, so that voltages $+\frac{1}{2}V_0$ and $-\frac{1}{2}V_0$ are applied to the fixed plates. Therefore the force acting on the plate on the balance arm will be

$$F = \frac{C^2}{2\varepsilon_0^2 A} [(V + \tfrac{1}{2}V_0)^2 - (V - \tfrac{1}{2}V_0)^2] = \frac{C^2 V_0}{\varepsilon_0^2 A} V, \tag{4.69}$$

giving a linear relation between the force and the applied voltage. Another arrangement sometimes used is similar to that of the quadrant electrometer.

As shown in Fig. 4.14, the beam of the torsion balance is in the shape of a paddle and is placed between two sets of plates, each in the form of four equal quadrants, which are connected to form two capacitors: C_{AC} between the torsion beam and A plates and C_{BC} between the torsion beam and B plates.

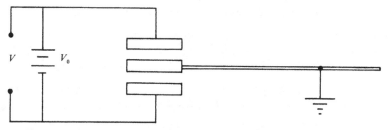

Fig. 4.13 Linearization of relation between force and applied voltage.

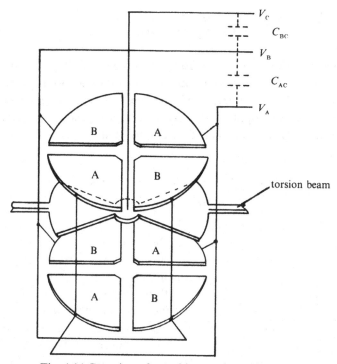

Fig. 4.14 Capacitors formed by quadrant electrometer.

The electrical energy of the two capacitors is given by

$$W = \tfrac{1}{2}C_{AC}(V_C - V_A)^2 + \tfrac{1}{2}C_{AB}(V_C - V_B)^2. \tag{4.70}$$

If θ is the angular displacement of the torsion balance relative to the fixed quadrant plates, then the electrical torque exerted on the torsion balance is given by

$$\tau = \frac{1}{2}\frac{\partial C_{AC}}{\partial \theta}(V_C - V_A)^2 + \frac{1}{2}\frac{\partial C_{BC}}{\partial \theta}(V_C - V_B)^2. \tag{4.71}$$

If $V_A = V_B$, there is no torque on the beam, and so

$$\frac{dC_{AC}}{d\theta} = \frac{dC_{BC}}{d\theta} = \frac{dC}{d\theta} = \text{const.} \qquad (4.72)$$

Hence,

$$\tau = \frac{dC}{d\theta}(V_A - V_B)[V_C - \tfrac{1}{2}(V_A + V_B)]. \qquad (4.73)$$

Thus, the torque τ can be found from the voltages V_A, V_B and V_C. It is convenient to take V_C to be much greater than V_A and V_B, so that

$$\tau = \frac{dC}{d\theta}V_C(V_A - V_B). \qquad (4.74)$$

For a constant value of V_C, the torque is proportional to $(V_A - V_B)$. The quadrant scheme could be used in both open- and closed-loop modes. In open loop, $(V_A - V_B)$ is the output of the meter due to the rotation of the pendulum; in closed loop, $(V_A - V_B)$ is the voltage applied to the meter to keep the torsion balance in its equilibrium position.

4.10 Torsion balance experiments

Terrestrial experiments

Eötvös began his experiments in 1890 (Eötvös, 1891) but his work extended over the next fifteen years and Renner came to repeat much of the work. A final account of Eötvös's results was only published three years after his death. In the Eötvös experiments, a single-beam torsion balance was used. The beam was 0.4 m long with masses of 30 g at the ends. One mass was fixed directly to the end of the arm, the other was suspended 212 mm below the opposite end. The suspension fibre was about 0.8 m long with a torsion constant of 0.5×10^{-7} N m rad^{-1}. The arm of the balance carried a mirror and a deflection of the arm was observed with a telescope viewing an image of a scale in the mirror. The arm length of the optical lever was about 0.6 m.

The beam and suspension were in a vessel with air at normal pressure and at room temperature. There were three metal walls with air between them which equalized the temperature across the enclosure to reduce convection current around the beam. The air also damped the oscillations of the beam, so that a steady reading was obtained about an hour or so after adjusting the apparatus.

The mass fixed directly to the beam was kept the same in a series of experiments, but the suspended mass was changed. The torque on the balance was measured by turning the whole balance in its enclosure through 180°, when the difference of deflection would correspond to twice the torque of eqn (4.55).

Renner (1935) made a further observation with some improvements to the apparatus and conditions.

Roll, Krotkov & Dicke (1964) raised two problems relating to the results of the experiments of Eötvös *et al.* (1922, 1935) and of Renner (1935). One problem was why the gravitational attraction of the observer at the telescope did not grossly disturb the result; Roll *et al.* conjectured that the observations were made so quickly that the balance did not have time to respond to the attraction of the observer. The other problem was the uncertainty of the data, but it is not necessary to repeat all the discussions here. Roll *et al.* (1964) concluded that the final results of Eötvös and Renner differed from those that were published. The results are: at 95 per cent confidence η is less than 9×10^{-9} for any combination of materials according to the experiments of Eötvös, while according to Renner, the corresponding limit is 4.2×10^{-9} (see Cook (1988a) for a discussion of the statistics of the result, and other matters).

Solar experiments

Eötvös (1891) was the first to use the torsion balance for tests on the weak principle of equivalence and although his terrestrial experiment is the best known, he also carried out a solar experiment using a single-beam torsion balance to look for a difference between magnalium and platinum. A platinum mass was attached to the south-pointing end, and hourly observations of the beam's orientation were made for two periods of two weeks, the first with a platinum mass on the north-pointing end and the second with a mass of magnalium. The value of η was calculated from the observations around sunrise and sunset, rather than from a Fourier analysis of all the observations. The result for platinum and magnalium in this first solar experiment was

$$\eta = 6 \times 10^{-9}.$$

No uncertainty was assigned.

Roll *et al.* (1964) performed another solar experiment using a triangular framework supporting three masses, one of gold and two of aluminium. The test mass of gold was a cylinder 32 mm long and 7.8 mm in diameter and the test masses of aluminium were cylinders 32 mm long and 21 mm in diameter, all weighing 30 g. The triangular framework from which they were suspended was of fused silica and the distance of each mass from the axis was 33 mm. In the initial experiments (with masses of lead chloride and copper instead of gold and aluminium) the suspension fibre was a tungsten ribbon with torsion constant of 7×10^{-9} N m rad^{-1}, but in the final form of the apparatus a fibre of fused silica was used, with a somewhat larger torsion constant of 2.4×10^{-8} N m rad^{-1} and a breaking strength of 2 N. The balance was enclosed in a stainless steel vessel with metallic seals, which was pumped out and then left sealed. In the preliminary work the pressure was 10^{-4} Pa at the end of the observations, while in the final apparatus it was 10^{-6} Pa at the end. Oscillations of the beam were damped with eddy currents. The upper end of the suspension fibre was fixed to a screw fitment which provided up and down movement and rotation, both controlled by discs of soft iron moved from outside the vacuum enclosure by magnets.

The rotation of the triangular frame was observed with an optical lever. The light from a small tungsten filament lamp was chopped by a wire oscillating at 3 kHz. The signal obtained from a photomultiplier was amplified by a phase-sensitive detector to give the displacement signal. That signal controlled a voltage applied to electrodes by which a torque was used to return the frame to the equilibrium position. Consideration of the radiation pressure and heating by the light source led to the 6 V filament lamp being run at 5 V from a stabilized voltage supply; not only were the possible disturbances greatly reduced but also the lamp lasted for a much longer time, which was important since the whole apparatus was left sealed. There was a steady drift of the suspension fibre and an electrostatic torque was needed to return the frame to the equilibrium position, and consequently a potentiometer, manually adjusted, was used to compensate for it. The whole apparatus was placed in a well-insulated pit, and thermocouples and thermistors were provided to measure temperature in the pit generally as well as close to the vacuum vessel.

In the design of the experiment, great attention was paid to a number of possible sources of error. The effect of variable gradients of gravity was reduced by remote operation, as well as by making the moment arms short (33 mm). Because the fundamental period of the experiment is 24 hours, the diurnal variation of the Earth's magnetic field would cause a spurious effect, so very careful precautions were taken to prevent all magnetic influence.

The initial experiment with lead chloride and copper using the tungsten fibre was carried out under good conditions. Long observations were made so that the data could be submitted to Fourier analysis. Signals of the expected 24 hour period were not detected, but others at 12 and 27.5 hour periods were found, thought to have been caused by changes of temperature and its rate. In this experiment the amplitude of the 24 hour period was found to be

$$\eta = (0 \pm 1.6) \times 10^{-10}.$$

When the final observations came to be made, construction work was underway near the experimental pit, disturbances from which interrupted the runs of observations. Most of the results were not long enough for a Fourier analysis. It was, however, possible to make a regression of the observed torque against the rates of change of temperature as measured by two of the thermistors placed close to the beam. The whole apparatus was rotated through 180° between runs and the most consistent values for η were obtained when the regression against the temperature measurement was subtracted. It was concluded that, with 95 per cent confidence, the Eötvös coefficient was less than 3×10^{-11} for gold and aluminium.

The discussion by Roll and others of their own work and also of that of Eötvös is extremely thorough, and should be read by anyone contemplating experiments on gravitation.

The solar experiment by Braginsky & Panov (1972) differed in a number of points of design and technique from the previous experiments. The suspended system was an eightfold symmetric framework with four masses of platinum and four of aluminium; the eight masses together weighed 3.9 g, and the beam assembly, as a whole, 4.4 g. The system was suspended by a tungsten fibre 5 μm in diameter and 290 mm long, which was pre-annealed by electric current heating to reduce the drift of the orientation of the beam assembly to 4×10^{-6} rad day^{-1}. The framework was in a glass vacuum vessel at a pressure of less than 10^{-6} Pa and was thoroughly shielded against magnetic and thermal disturbance.

Light from a laser was reflected from a mirror at the centre of the framework onto a photographic film on a rotating drum. To amplify the excursions of the reflected beam on the surface of the drum, the axis of the drum was set to make an angle of 0.2 rad to the reflected light; according to the study of section 4.9 this will increase the power of the amplification of the optical lever by a factor of 5. The amplitude of the 24 hour component of the rotation of the balance was measured directly from the film. The duration of a single record was rather short, the longest being 3 days, so that only two or three 24 hour oscillations occurred in any run. Altogether, seven runs were recorded, as shown in Table 4.2. From these

Table 4.2. *Results of Braginsky & Panov (1972)*

Date in Feb. 1971	Deflection (10^{-7} rad)
1–4	+2.35
8–10	−0.40
10–12	−1.70
12, 13	+0.76
14, 15	−2.96
15–17	−0.10
19–21	−1.77

data they concluded that at 90 per cent confidence the Eötvös coefficient is less than 9×10^{-13} for platinum and aluminium.

In summary, the above experiments are so far the most delicate mechanical measurements undertaken and owe their success to careful design, sensitive apparatus and thorough tests and observation. However, it seems clear that yet more delicate observations could be made. The work of Eötvös and his successors, advanced and careful as it was in its day, was limited by the available technologies of observation; the apparatus of Roll *et al.* (1964) in its final form was not fully exploited because of disturbance from building works; and the observations of Braginsky & Panov (1972) lasted for periods of only a few days.

4.11 Experiments with beam balance

Until recently it was generally accepted that the torsion balance was the most sensitive detector of mechanical forces, but recent work (Speake, 1987; Speake & Quinn, 1987) has shown that similar sensitivities should be attained with a beam balance and consequently that it could be used in tests of the weak principle of equivalence. The beam balance is the traditional device for making direct mass comparisons. Modern beam balances may have a reproducibility better than 1 part in 10^9 and a resolution of 10^{-10} m s^{-2} (see Speake & Gillies, 1987) and so, prompted by the search for the new hypercharge force predicted by Fischbach *et al.* (1986), Speake & Quinn (1987, 1988) proposed a new experiment using a beam balance. This experiment could also be interpreted as comparing the inertial mass and gravitational mass of the test body.

According to their proposal, a transportable beam balance with two test bodies of different materials would be used to compare the masses at two

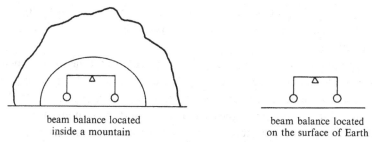

beam balance located
inside a mountain

beam balance located
on the surface of Earth

Fig. 4.15 Test of weak principle of equivalence with sensitive balance used at
different sites.

different sites, the first in a tunnel deep inside a mountain with several
kilometres of cover. Here, the test bodies would be adjusted so as to have
apparently equal masses as measured using the beam balance. The whole
apparatus would then be moved out of the tunnel to a second site on the
surface of the Earth, preferably at the same altitude and not too distant.
The measurement would be repeated to find whether the balance changed.
A difference of the apparent weight Δw is the measure of the Eötvös
coefficient for the two materials used in the experiment.

In the first site, as shown in Fig. 4.15, the equation for the balance of two
test bodies a and b is given by

$$m_{ag} g_1 - m_{ai} R\omega^2 \cos^2 \lambda = m_{bg} g_1 - m_{bi} R\omega^2 \cos^2 \lambda, \qquad (4.75)$$

where the m_{ag}, m_{bg} and m_{ai}, m_{bi} are the passive gravitational masses and
inertial masses of test bodies a and b respectively, ω is the spin frequency
of the Earth, λ is the latitude of the first site, R is the distance of the beam
balance to the centre of the Earth, g_1 is the local value of gravity inside the
mountain. If the second site is located at the same latitude and altitude, the
equation for the balance at the second site is then

$$m_{ag} g_2 - m_{ai} R\omega^2 \cos^2 \lambda = m_{bg} g_2 - m_{bi} R\omega^2 \cos^2 \lambda + \Delta w, \qquad (4.76)$$

where g_2 is the local value of gravity on the surface of the Earth, which
differs from g_1 in the first site. Combining the above two equations, it is
found that

$$(m_{ag} - m_{bg})(g_2 - g_1) = \Delta w. \qquad (4.77)$$

Since the inertial masses m_{ai} and m_{bi} can only differ slightly, we take m be
the average value of these two, which is very close to m_{ai} and m_{bi}. If eqn
(4.77) is divided by m, then the Eötvös coefficient is given by

$$\eta = \frac{\Delta w}{mg} \frac{g}{g_2 - g_1}. \qquad (4.78)$$

$\Delta w/mg$ can be 10^{-9} or better in a highly stable balance, while $g/(g_2 - g_1)$ may not exceed 100. Thus it should be possible to set a limit of about 10^{-7} to 10^{-8} for η.

Although the precision of a beam balance used to test the weak principle of equivalence may be not as high as that of the torsion balance, it is still quite essential to employ it. It is important to use a variety of methods to test such an important principle even if all methods cannot give the highest precision. Thus in experiments using a torsion balance the test bodies are stationary with respect to the Earth and to each other, whereas in experiments using the free fall method (section 4.6), in which the precision approaches that of the torsion balance, the test bodies have relative velocities with respect to the Earth and each other. Another point is that, in all previous high precision experiments, the mass of the test body ranges from a few grams to a few tens of grams, but in experiments carried out by the beam balance it would be easy to extend the range of mass to kilograms or more.

5

Verification of the weak principle of equivalence for free particles

5.1 Introduction

Although the weak principle of equivalence has been verified for ordinary macroscopic matter to very high precision, two questions remain open:

(1) Is the principle valid for antimatter? Although indirect evidence from virtual antimatter in nuclei and short-lived antiparticles suggests that antimatter may have normal gravitational properties, no direct tests of the validity of the weak principle of equivalence for antimatter have been made.

(2) Is the principle valid for microparticles? As the test bodies in macroscopic experiments are formed of neutrons, protons and electrons bound in nuclei, there is no doubt about the validity of the weak principle of equivalence for bound particles. However, the possibility of the principle of equivalence being violated for free particles should be studied.

Two main features characterize laboratory tests of the weak principle of equivalence for free elementary particles, both the consequence of their small masses. (1) When forces on substantial masses of bulk material are compared, a null experiment based on comparing different test bodies of two kinds of material can be devised. That is not possible for microscopic particles, and the gravitational accelerations have to be measured directly and subsequently compared with the acceleration of ordinary bulk matter to obtain the Eötvös coefficient. (2) The gravitational forces are very weak, even in the field of the Earth (which is the strongest attractive field), and so the accuracy of any experiment is very poor compared with Eötvös-type experiments using bulk masses.

The movement of particles is governed by quantum mechanics, and the

97

inertial movement is related to Planck's constant h; the law of attraction is governed by gravitational theory, and this force is related to the gravitational constant G. The results of the experiments thus involve the ratio h/G, and the experiments, although of relatively low precision, are important as they may indicate whether quantum laws are affected by gravity, or they may give direct evidence about possible gravitational interactions with other fundamental forces of nature – the electromagnetic force, the weak force and strong force – and so are relevant to the possible establishment of a future unification theory.

Experiments have been performed on atoms, neutrons, electrons and photons, as discussed in the following sections.

5.2 Experiments on atoms

The validity of the weak principle of equivalence for atoms was first tested by Estermann, Simpson & Stern (1947) when they measured the velocity distribution of molecular beams of potassium and caesium atoms.

A thermal beam of atoms produced from an oven will have a certain velocity distribution which will make the intensity distribution take a certain form for a horizontal beam deflected by the vertical gravitational attraction of the Earth, as shown in Fig. 5.1.

Although those experiments were devised to verify the modified Maxwellian velocity distribution within the beams, they also show that the free atoms follow the laws of free fall in the same way as bulk matter – direct evidence that atoms obey the weak principle of equivalence.

As shown in Fig. 5.1, an atom leaving an oven with velocity v at an angle α above the horizontal plane will fall on a detector at a distance $s(v)$ below the plane of slit and oven. The intensity distribution in the plane of the detector in Fig. 5.1 can thus give the value g_a, for the acceleration of atoms in the field of the Earth. A comparison of g_a with g, the acceleration for bulk matter, yields the Eötvös coefficient for atoms.

Let the distance between the oven and the slit be L_1, and that between the slit and the detector be L_2. Then

$$L_1 = \frac{v^2 \sin 2\alpha}{g}. \tag{5.1}$$

Hence particles with velocity v strike the detector at a distance $s(v)$ below the plane of the slit and oven give by

$$s(v) = L_2 \tan \alpha + \frac{1}{2} \frac{g_a L_2^2}{v^2} \sec^2 \alpha. \tag{5.2}$$

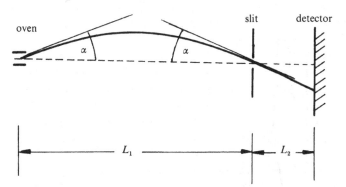

Fig. 5.1 Freely falling atoms in test of weak principle of equivalence.

Now α is very small because the thermal velocity v is large compared with the velocity acquired in falling in the Earth's gravitational field and the distance L_1 is much greater than the opening of the oven, that is, gL_1/v^2 is small in any practical apparatus. In that case,

$$\sin \alpha = \frac{1}{2}\frac{g_a L_1}{v^2} \tag{5.3}$$

and

$$s(v) = \frac{1}{2}\frac{g_a L_2}{v^2}(L_1 + L_2) \tag{5.4}$$

or

$$g_a = \frac{2v^2 s(v)}{(L_1 + L_2)L_2}. \tag{5.5}$$

In the above derivation, it is assumed that the atoms obey the weak principle of equivalence, that is why there are no terms of m_i/m_g appearing in the formula. However, whether the assumption is right or not will be shown by the comparison of g_a to g, the local gravity for bulk material, which gives for the Eötvös coefficient:

$$\eta = 1 - \frac{g_a}{g}. \tag{5.6}$$

A measurement of the gravitational accelerations g_a of the free atoms requires values of both the velocities of the atoms and the displacement $s(v)$. Since the velocities of thermal atoms follow the Maxwellian distribution, the intensities of the atoms falling on D at different distances s can be calculated and a comparison of the experimental and theoretical

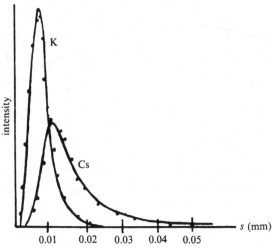

Fig. 5.2 Gravitational deflection of atoms of caesium and potassium (Estermann *et al.*, 1947), dots: experimental results; line: theory.

distribution will test the validity of the weak principle of equivalence. The Maxwellian distribution of the number of atoms at distance s is given by

$$n(s) \, \mathrm{d}s = n_0 \, s \exp\left(\frac{s_0}{s}\right) \mathrm{d}s, \qquad (5.7)$$

where

$$s_0 = \frac{m}{2k_b T} g_a (L_1 + L_2) L_2. \qquad (5.8)$$

The maximum of the distribution occurs at $s = s_0$. Thus

$$g_a = \frac{2k_b T s_0}{m(L_1 + L_2) L_2}. \qquad (5.9)$$

The determination of g_a requires a value of the effective temperature T or, alternatively, a direct measurement of the velocity. In the experiments of Estermann *et al.*, L_1 and L_2 were both equal to 1 m; the comparison of the experimental results and the calculated results for caesium atoms and potassium atoms is shown in Fig. 5.2.

The experiment demonstrated a gravitational acceleration of free caesium and potassium atoms in an atomic beam which apparently agreed to within a small percentage with that of bulk matter. Although the precision is not great, it is a valuable practical test of the weak principle of equivalence. Following the same line, later, more accurate results were obtained with neutrons.

5.3 Experiments on neutrons

The effective gravitational acceleration of neutrons may also be found from the drop of a collimated beam, as was first done by McReynolds (1951) using a highly collimated beam of thermal neutrons from the reactor of the Brookhaven National Laboratory. The flight path was 11.6 m long, and instead of measuring the intensity distribution for a single beam with a spread of velocities, two beams with different known velocities were used. One beam was the normal thermal neutron beam (fast beam), the other one was filtered by a 25 cm filter of beryllium oxide interposed between the reactor and first slit and passing only very slow neutrons. A value of g_n for neutrons can be calculated from the separation of the two free-fall distributions. In McReynolds's experiment, a downward shift of the slower neutron beam of 1.22 mm in the 11.6 m path was observed. According to eqn (5.5), the gravitational acceleration of the neutrons is given by

$$g_n = \frac{2\Delta}{L_1 L_2 + L_2^2} \frac{1}{\overline{v_1^{-2}} - \overline{v_2^{-2}}}, \tag{5.10}$$

where $\overline{v_1^{-2}}$ and $\overline{v_2^{-2}}$ are the average values of the inverse square of the velocities of the two beams and Δ is their separation at the detector. McReynolds found that

$$g_n = 9.35 \pm 0.70 \text{ m s}^{-2},$$

leading to

$$\eta = 4 \times 10^{-2}$$

for the difference between neutrons and bulk matter.

McReynolds achieved better precision than Estermann et al. (1947) because he used two beams, fast and slow: with just one beam it is difficult to define the unaffected position. Later experimenters adopted the two-beam method. However, in McReynolds's experiments, the slower velocities were not very well defined, as the beryllium oxide filter does not have a sharp cut-off. To obtain better precision, Dabbs et al. (1965) used a crystal filter to define the velocities of neutrons in the slow beams.

The velocities of neutrons correspond to de Broglie wavelengths given by

$$\lambda = \frac{h}{mv}, \tag{5.11}$$

where h is Planck's constant, and m is the mass of the neutron.

Dabbs et al. (1965) used the transmission edges associated with the

$\langle 001 \rangle$ and $\langle 002 \rangle$ lattice spacing in a polycrystalline beryllium filter to determine the wavelengths, and hence the velocities of the slower neutrons. This method of filtering not only gives sharply distinguishable velocities, it gives high intensity as well. The filter is located just in front of the detector and near the end of the flight path (which was much longer (180 m) than in the earlier experiments). The detectors were BF3 proportional counters.

The lattice spacings of beryllium are

$$2d_{100} = 3.9558 \pm 0.0016 \,\text{Å},$$
$$2d_{002} = 3.5780 \pm 0.0016 \,\text{Å}.$$

They define the velocities of two slower neutron beams as 1.0×10^5 cm s^{-1} and 1.106×10^5 cm s^{-1} respectively. The deviations of the two slower beams were compared with that of a fast beam which did not pass any filter and did not fall appreciably. The gravitational acceleration of neutrons, as calculated for each of the two slow beams for a path 180 m long, was as follows:

lattice spacing $\langle 002 \rangle$, deviation 15.5 cm, $g_n = 9.731 \pm 0.074$ m s^{-2},

and

lattice spacing $\langle 001 \rangle$, deviation 12.7 cm, $g_n = 9.754 \pm 0.031$ m s^{-2}.

The local value of g is 9.7974 m s^{-2}, so that the Eötvös coefficients for neutrons were found to be

$$\eta_1 < 7 \times 10^{-3},$$
$$\eta_2 < 4 \times 10^{-3},$$

a very good precision for microparticles. Dabbs *et al.* attained such a relatively high precision because they used the lattice spacing of a crystal to determine the velocities and because they had a much longer flight path (180 m, compared to 1 m in the experiment of Estermann and 11.6 m in the experiment of McReynolds) and thus much greater deviation of the beams.

Errors in this experiment may come, among others, from the errors in the lattice spacing, a deviation of fast beam drop caused by Coriolis force and variations in the scattering cross-section of beryllium.

The above experiments are based on the comparison of two neutron beams; however, a quite different approach was adopted by Koester (1976). Instead of, in effect, measuring the vertical velocity acquired by a particle in a time v/L_c, Koester compared the potential gained by a neutron in free fall over a distance h with the known potential of neutrons in a

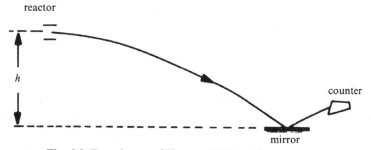

Fig. 5.3 Experiment of Koester (1976) with neutrons.

liquid; the latter can be found from transmission measurements on liquid probes (see Koester & Nistler, 1975; Koester, Nistler & Waschkowski, 1976; Waschkowski & Koester, 1976). This is known as a neutron gravity refractometer. Thus, Koester's experiment involves the measurement of the scattering length of a slow beam after a free fall in the field of gravity. This length then can be compared with the 'true' value of the scattering length, which can be measured independently of gravity.

In Koester's experiment neutrons from a reactor travel for about 100 m starting along a nearly horizontal direction and fall under gravity, through a distance h, so gaining kinetic energy $mg_n h$. They fall on a horizontal mirror consisting of thin layers, about 3 mm thick, of a pure liquid over a glass tray. Total internal reflection occurs when the kinetic energy associated with the vertical component of the momentum is equal to the mean potential of the neutrons inside the liquid. The total reflection is monitored by a counter as shown in Fig. 5.3.

The mean potential of the neutron inside the mirror substance can be expressed by

$$V = \frac{2\pi\hbar^2}{m} Na_c,$$ (5.12)

where m is the mass of the neutron, N is the number of atoms per cm^3, $\hbar$ is Planck's barred constant, and a_c is the scattering length which can be obtained from precise transmission experiments. Then, since the kinetic energy gained by the falling neutrons from the Earth's gravity is $mg_n h$, we have

$$g_n = \frac{2\pi\hbar^2}{m^2 h} Na_c.$$ (5.13)

Eighteen various liquids of eleven different organic substances containing the elements carbon, hydrogen and chlorine were used in the experiments, from the results of which, Koester concluded that, for free neutrons, the Eötvös coefficient is

$$\eta = 2.5 \times 10^{-4}.$$

5.4 Experiments on photons

Tests of the photon red shift predicted by the theory of general relativity are all tests of the weak principle of equivalence. The most successful experimental tests of the weak principle for photons in the laboratory are the gravitational experiments on gamma rays using the Mössbauer effect by Cranshaw & Schiffer (1964) and Pound & Snider (1965), in which the change in frequency of a gamma ray arising from a difference of gravitational potential is compensated by the Doppler shift under known velocity. It is one of the advantages of laboratory experiments that it is possible to use velocity compensation to measure the shift of frequency.

If a photon of frequency v_0 falls a vertical distance l in the field of the Earth it will gain energy $m_g gl$, where m_g is the passive gravitational mass of the photon, consequently

$$\hbar v_0 + m_g gl = \hbar v. \tag{5.14}$$

If the weak principle of equivalence holds for photons, the passive gravitational mass equals the inertial mass determined by $m_i = \hbar v/c^2$, and then

$$\frac{\Delta v}{v} = \frac{gl}{c^2}. \tag{5.15}$$

The relative change of frequency in a laboratory at the surface of the Earth is extremely small, only $1.1 \times 10^{-16} \text{ m}^{-1}$.

The resonance spectroscopy of gamma rays, known as Mössbauer spectroscopy, is a powerful tool to measure such small frequency shifts in the laboratory. Because the motions of nuclei in ^{57}Fe are correlated, the relative linewidth, that is, the ratio of the linewidth to the frequency of the emitted gamma rays, is essentially determined by the uncertainty principle, for the Doppler shift due to recoil velocity is very small; in ^{57}Fe it is 3.1×10^{-13}. On the other hand, if the height l is 22.5 m, from eqn (5.15) the frequency shift is about 2.5×10^{-15}, which is one part in 140 of the natural linewidth and is measurable.

To measure this small shift, Pound and Snider gave the iron source a periodic velocity and measured the dependence of absorption in a second sample as a function of the velocity of the source.

The experiments were performed in the Jefferson Physical Laboratory of Harvard University over a 22.5 m vertical interval. The source was of ^{57}Co, initially 1.25 curies. It was located on the top of the building, and was mounted on a ferroelectric transducer which gave it a harmonic vertical motion with a frequency of 73 Hz. The transducer was attached to a hydraulic piston which moved at a steady velocity of 7×10^{-6} m s^{-1}, corresponding to the gravitational shift of the frequency of the gamma rays of ^{57}Fe over 22.5 m. The resonant absorber was a set of iron foils enriched in ^{57}Fe and 38 cm in diameter. The vertical path was in a plastic tube 40 cm in diameter filled with helium to avoid absorption in air. The gamma radiation observed was that at 14.4 keV.

To avoid the systematic errors caused by the broadening of linewidth due to different thermal Doppler effects, a regulated oven kept the source, absorbers, and their associated proportional counters, at $43.5 \pm 5 \times 10^{-3}$ °C in the time of a given run. In addition, precision thermistors were used to record continually the temperature of the source and absorbers, so that corrections could be made to the final results.

The observed relative change of frequency was found to be (0.9990 ± 0.0076) times the expected value of 4.905 ± 10^{-15}. The Eötvös coefficient for gamma rays is then

$$\eta < 10^{-2}.$$

5.5 Experiments on free electrons

The weak principle of equivalence for free electrons may be tested by the motion of electrons in the gravitational field of the Earth in the same way as for atoms and neutrons, but it is much less easy because of the electrical charge of the electrons and the small mass. An alternative is to measure the force on electrons in free fall within an enclosed metallic shell.

A quantum mechanical argument shows (Schiff & Barnhill, 1966; Schiff, 1970) that if the weak principle of equivalence is valid for electrons and if they are located in the space of a closed metallic shell, they are subject to a uniform electric field E_w which equals mg/e, where m is the mass and e the charge of the electron. This field is directed so as to exert an upward force on an electron. Inside such a shield, free electrons will not fall under gravity, since the weight is exactly balanced by the electric field. On the other hand, positrons, if they do obey the weak principle of equivalence, should fall with acceleration of $2g$.

The most successful experiment to measure the force exerted on electrons inside a metallic shell, is that of Witteborn & Fairbank (1967) who

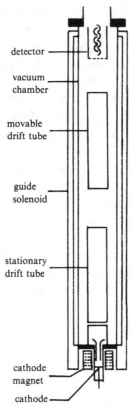

detector

vacuum
chamber

movable
drift tube

guide
solenoid

stationary
drift tube

cathode
magnet

cathode

Fig. 5.4 Experiment of Witteborn and Fairbank (1967) on electrons.

measured the time-of-flight distribution of electrons falling freely within a
metal enclosure. Within the enclosure, all vertical electric and magnetic
potential-energy gradients were reduced to below 10^{-11} eV m^{-1} except for
known and deliberately applied uniform electric fields. As shown in Fig.
5.4, electrons were emitted in a short burst from a cathode placed at the
bottom of a long (91 cm) vertical copper drift tube having a very uniform
inside diameter. The copper tube was located inside a vacuum chamber,
having the wall cooled to 4.2 K. The electrons were constrained to move
along the axis of the drift tube by an axial magnetic field produced by a
coaxial superconducting solenoid. Electrons which passed upwards
through the stationary drift tube and the movable drift tube were detected
by a 14-dynode electron multiplier detector which produced a measurable
voltage pulse each time an electron entered it. Pulses were amplified and
stored in a 400-channel analyser and the time of flight of electrons through
the drift tube was derived from the data. A weak uniform electric field E_a

was maintained inside the drift tube by passing a d.c. current through the walls of the drift tube parallel to its axis. Thus, an electron inside the tube will experience the force

$$f = mg_e - eE_w + eE_a,$$ (5.16)

where g_e is the gravitational acceleration of the electron in the field of the Earth. Then the maximum observable flight time through height h is

$$t_{max} = \left(\frac{2hm}{mg_e - eE_w + eE_a}\right)^{\frac{1}{2}},$$ (5.17)

which is a function of E_a.

A change of the externally applied field E_a will alter the time of the flight. After about 10 000 pulses were accumulated, the distribution of the time of flight could be analysed. The study of such distributions for several different applied fields E_a yielded the mass of the electron and the value of $(mg_e - eE_w)$. It was found that t_{max} varied inversely with the square root of E_a, which indicated that the value of $(mg_e - eE_w)$ must be extremely small. An average of 11 of the data sets gave

$$mg_e - eE_w \leqslant 0.51 \times 10^{-11} \text{ eV m}^{-1} = 0.09 \ mg.$$ (5.18)

Thus the gravitational acceleration of the electron is

$$g_e = g(1 \pm 0.09),$$ (5.19)

and the experiments of Witteborn and Fairbank set a limit for the Eötvös coefficient for the free electron of

$$\eta < 0.09.$$ (5.20)

5.6 Conclusion

The foregoing brief review of the principles and results of the weak principle of equivalence for elementary particles shows that not only is it desirable to improve the precision, but also that further experiments should be conducted on a wider range of particles, particularly antimatter particles. The best discrimination of experiments on elementary particles, about 1 in 10^4 for neutrons, is comparable with the best results of laboratory tests of the inverse square law and, as pointed out in the Introduction (Chapter 1), far inferior to that of tests of weak equivalence on bulk matter. In current quantum-gravity theories, the graviton has spin-1 and spin-0 partners, the effects of which produce measurable non-

Newtonian phenomena, breaking the principle of equivalence. The direct measurements of the gravitational acceleration of antimatter particles could prove the existence of these new partners. Although several experiments are currently in progress (see, Goldman, Hughes & Nieto, 1987), no confirmed results have been published.

It should be noted that since bulk matter is composed essentially of neutrons and protons, and since the weak principle of equivalence is very well satisfied for bulk matter, it must be satisfied by protons as it is found to be for neutrons.

6

Newtonian attractions of extended bodies

6.1 Introduction

In tests of the weak principle of equivalence, exact calculations of the attractions of masses are not necessary, but they are essential in experiments to test the inverse square law and to measure the gravitational constant. In fact, the calculation of the gravitational attraction of laboratory masses is usually not at all simple, because the dimensions of the masses are comparable with the separations between them, so that neither the test mass nor the attracting mass can be treated as a point object. In the following sections we discuss the gravitational attractions of laboratory masses with various common geometrical shapes. We present the results in terms of the gravitational efficiency, that is, the ratio of the gravitational attraction of a laboratory mass at a certain separation to that of a point mass with the same mass and separation. Furthermore, the precision demanded in measurements of separations of masses, the most difficult measurements in the determination of G and the test of gravitational law, depends on the geometry of the masses. These effects can have a strong influence on the conduct and final results of an experiment and it is essential to discuss in detail the calculation of potentials and attractions before going on to describe experiments.

Masses of three forms are often used in the laboratory: spheres, cylinders and rectangular prisms. The formula for the gravitational attraction of a sphere is well known and simple, but in practice it is not possible to manufacture an ideal sphere, the practical problem is usually how the real precision of manufacture affects the results; cylinders and prisms can be made very precisely but calculating the attraction is difficult. All three forms will be discussed below. Naturally the results are applicable both to attracting and to attracted masses. It can be expected that two sorts

109

of error can cause deviation from ideal conditions, on the one hand the form of the surface may depart from the ideal geometry, and on the other, there may be irregularities of density within the body. Both types of error are discussed below.

6.2 Attraction of a sphere

We consider bodies that are nominally but not perfectly spherical and ask by how much the attraction departs from that of a perfect sphere.

Real imperfect spheres deviate from a perfect one in two ways: (1) the density distribution is inhomogeneous, and (2) the geometry deviates from a spherical form.

Effect of inhomogeneous density distribution

We take spherical polar coordinates (r, θ, ϕ) and suppose that the density of a sphere is a function of position, $\rho(r, \theta, \phi)$. The gravitational potential at an external point in the direction $\phi = 0$ (the polar axis) is (see Fig. 6.1)

$$V = -G \int_0^\pi \int_0^{2\pi} \int_0^{r_0} \frac{\rho(r, \theta, \phi) \, r^2 \sin\phi \, d\phi \, d\theta \, dr}{(R^2 + r^2 - 2Rr\cos\phi)^{\frac{1}{2}}}. \tag{6.1}$$

The formula is quite general on account of spherical symmetry – any axis may be taken as the polar axis but ρ will vary with the choice.

The general expression for the external potential of any body is

$$V = -\frac{GM}{R} + G \sum_n \left(\frac{a}{R}\right)^n \sum_m P_n^m(\cos\theta)(J_{nmc}\cos m\phi + J_{nms}\sin m\phi), \tag{6.2}$$

where a is the radius of a sphere containing all the matter, J_{nmc} and J_{nms} are multipole moments of the density and $P_n^m(\cos\theta)$ is an associated Legendre function. The calculation of the deviation from $-GM/R$ is difficult since the distribution of density is unknown in general but the error can be eliminated provided the inhomogeneity of the density is not too great, as will usually be so in gravitational experiments. The experiment should be designed so that the sphere can be rotated about three mutually perpendicular axes; the final value of the acceleration is obtained by averaging many measurements, with many orientations of the sphere. The average attraction will be equal to that of a point mass M at the centre of the sphere. The rotations average out variations of density depending on

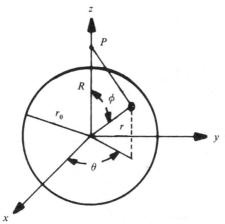

Fig. 6.1 Coordinates in sphere of inhomogeneous density.

direction while inhomogeneity in the radial direction remains. However, that inhomogeneity does not affect the form of the gravitational attraction at a point outside the sphere.

The effect of taking the mean as above is to replace the function $\rho(r, \theta, \phi)$ by $\rho(r)$, since only the mean radial inhomogeneity remains. Then on account of the integral relation

$$\int_0^\pi \frac{\sin \phi \, \mathrm{d}\phi}{(R^2 + r^2 - 2Rr \cos \phi)^{\frac{1}{2}}} = \frac{2}{R}, \tag{6.3}$$

eqn (6.1) becomes

$$V = -\frac{G}{R} \int_0^{r_0} 4\pi r^2 \rho(r) \, \mathrm{d}r = -\frac{GM}{R}, \tag{6.4}$$

which is the well-known result that the potential of any set of concentric spherical shells is equal to $-GM/R$, where M is the total mass of the shells.

For the averaging procedure to be effective the number of orientations should be large, the angular size of the rotation should be small.

Effect of departures from a perfect sphere

Here, we consider a laboratory sphere that has a homogeneous density distribution, but with an imperfect geometrical shape. Again we will deal with the case in which the geometrical departure from the perfect spherical shape is not very large. To derive the appropriate formula for the attraction, we make two changes to eqn (6.1) which gave the gravitational

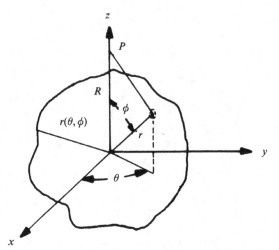

Fig. 6.2 Sphere of irregular form.

potential for a sphere with an inhomogeneous density distribution. First, the density ρ is constant, instead of being a function of r, θ and ϕ; and secondly, as shown in Fig. 6.2, the integral limit for the variable r is a function $r(\theta, \phi)$, instead of being a constant. Therefore eqn (6.1) will be replaced by

$$V = -G \int_0^\pi \int_0^{2\pi} \int_0^{r(\theta, \phi)} \frac{\rho r^2 \sin\phi \, d\phi \, d\theta \, dr}{(R^2 + r^2 - 2Rr\cos\phi)^{\frac{1}{2}}}. \tag{6.5}$$

If a body deviates only slightly from a perfect sphere, it will be sufficient to take the potential in the form.

$$V = -\frac{GM}{R}\left(1 + \frac{A + B + C - 3I}{2MR^2} + \cdots\right), \tag{6.6}$$

where A, B and C are respectively the least, intermediate and greatest moments of inertia of the body and I is the moment about the z-axis. Furthermore, the attraction produced by a geometrically irregular laboratory sphere upon the point P on the z-axis is

$$\text{acceleration} = \frac{GM}{R^2}\left[1 + \frac{3(A + B + C - 3I)}{2MR^2} + \cdots\right]. \tag{6.7}$$

We conclude that measurements of the moments of inertia about three orthogonal axes should be made in order to evaluate the effect of the non-spherical part of the attracting mass. While that needs great care it is

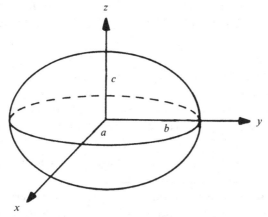

Fig. 6.3 Parameters of an ellipsoid.

nevertheless easier than a point-to-point measurement of the surface of the mass as required by the integration in eqn (6.5). In some special cases, the procedure can be made even simpler.

For a laboratory sphere, of all the possible slightly irregular shapes, the ellipsoid is the most common one. As shown in Fig. 6.3, the moments of inertia of an ellipsoid about x, y and z axes are given by

$$A = \tfrac{1}{5} M(b^2 + c^2),$$
$$B = \tfrac{1}{5} M(a^2 + c^2),$$
$$C = \tfrac{1}{5} M(a^2 + b^2),$$

respectively, where a, b and c are the radii of the ellipsoid along x, y and z axes. Therefore, the deviation of the acceleration at a point P on the z-axis of an ellipsoid from that produced by a perfect sphere is estimated as

$$\Delta \equiv \text{acceleration} - \frac{GM}{R^2} \approx \left[\frac{3(2c^2 - a^2 - b^2)}{10R^2} \right] \frac{GM}{R^2}. \tag{6.8}$$

Eqn (6.8) indicates that it is not necessary to measure the moments of inertia A, B and C; it is only necessary to measure the diameters of the sphere along three orthogonal axes.

Although the above expression shows that the deviation will decrease as R (the distance between P and the centre of the laboratory sphere) increases, in practice, in order to obtain a large gravitational acceleration, R should be kept as small as possible, which means that the deviation in eqn (6.8) cannot be significantly reduced by increasing R.

We may estimate the effect with some typical values. Consider a precision ball with dimensions $a = b \approx c = 50$ mm; R, the separation between the attracting and attracted spheres, may be assumed to be 100 mm; the best possible manufacture of the ball gives a deviation from sphericity of about 12.5 μm, which may be expressed as $c - a \approx 6$ μm. For these figures, the deviation Δ in eqn (6.18) becomes 7×10^{-5}, which makes a large contribution to the final error of G determined by an experiment using such a sphere as the attracting mass. Therefore, we conclude that in experiments where a sphere is used as the attracting mass, a very high precision ball is needed to prevent systematic error.

In the above text the effects of inhomogeneous density and irregular shape were considered separately. A real laboratory ball, however, suffers from both those imperfections; the overall effect may be very complicated, but if the two imperfections are small enough they may be estimated independently.

6.3 Attractions of cylinders

The use of a sphere in gravitational experiments has some advantages: the calculation is simple, the gravitational efficiency is a maximum, equal to one, but the preparation of a high-precision sphere is difficult and costly. A right circular cylinder is much easier and less expensive to manufacture.

Generally speaking, the regularity of a cylinder can be one or two orders better than that of a sphere. For an example, for a precision ball of 100 mm diameter, the deviation from sphericity usually is about 10 μm, or about 1 part in 10^4. For a well-made cylinder of 100 mm diameter, the deviation from circularity may be less than 1 μm even using an ordinary lathe, that is 1 part in 10^5 and much better than that of a sphere. Indeed the manufacturing precision of a cylinder is well beyond the requirements of gravitational experiments, although variations of density may be significant.

The gravitational attraction of a finite cylinder is not easy to calculate. Despite a number of discussions (Basset, 1886; Poynting, 1891; Heyl, 1930; Heyl & Chrzanowski, 1942; MacMillan, 1958), no closed analytical result has been given. Heyl (1930) published the result of his calculation of the radial gravitational field of a cylinder in the form of an extensive polynomial covering nearly ten pages but with a precision of not better than 10^{-6} (Cook & Chen, 1982).

We now obtain a closed analytical result for the gravitational attraction of a finite cylinder at an arbitrary point. The derivation is lengthy and tedious, and only the key steps will be described (for the details, see Cook & Chen, 1982).

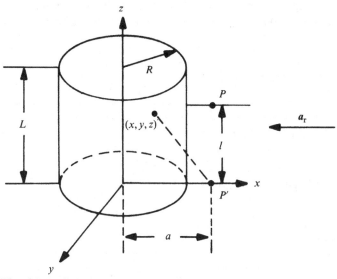

Fig. 6.4 Radial gravitational attraction of a right circular cylinder.

The attraction of a finite cylinder at an arbitrary point may be resolved into axial and radial components:

$$\text{acceleration} = a_a + a_r, \tag{6.9}$$

where a_a is the acceleration upon a point P in the axial direction and a_r is that in the radial direction. They are discussed separately.

Radial gravitational attraction of a finite cylinder
The $C_y(L, R, a)$ function

It will be shown here that the radial attraction of a cylinder can be written in terms of a function, denoted by $C_y(L, R, a)$, of the dimensions of the cylinder and the distance of a field point.

Consider a solid cylinder (Fig. 6.4) of length L, radius R and density ρ (assumed to be a constant). We consider first the attraction at a point P' in the basal plane of the cylinder. The result may be generalized for any point P by superposing attractions of component cylinders.

A Cartesian coordinate system is used in the calculation with P and P' located in the (x, z)-plane and the z-axis coincident with the axis of the cylinder, as shown (Fig. 6.4). By axial symmetry, the choice of coordinates and the results will apply to a general point P' in the (x, y)-plane.

Let a be the radial distance of the point P' from the axis of the cylinder, so that the magnitude of the radial gravitational attraction a'_r due to the solid cylinder at the point P', which is the radial differential of the potential, is

$$a'_r = \frac{\partial}{\partial a}\left\{-2\int_{-R}^{R}dy\int_{-(R^2-y^2)^{\frac{1}{2}}}^{(R^2-y^2)^{\frac{1}{2}}}dx\int_0^L\frac{G\rho}{[y^2+z^2+(x-a)^2]^{\frac{1}{2}}}dz\right\}. \quad (6.10)$$

On differentiating with respect to the variable a and integrating with respect to z and x, we find

$$a'_r = 2G\rho(-I_1+I_2), \quad (6.11)$$

where I_1 and I_2 are the integrals

$$I_1 = \int_0^R\ln\left\{\frac{L+[L^2+R^2+a^2+2a(R^2-y^2)^{\frac{1}{2}}]^{\frac{1}{2}}}{L+[L^2+R^2+a^2-2a(R^2-y^2)^{\frac{1}{2}}]^{\frac{1}{2}}}\right\}dy, \quad (6.12)$$

$$I_2 = \frac{1}{2}\int_0^R\ln\left[\frac{R^2+a^2+2a(R^2-y^2)^{\frac{1}{2}}}{R^2+a^2-2a(R^2-y^2)^{\frac{1}{2}}}\right]dy. \quad (6.13)$$

The evaluation of I_2 is relatively easy, but that of I_1 is not so straightforward. A dummy variable transformation is used as a means of obtaining the integral as a function of a third variable with respect to which I_2 can be evaluated, but which does not appear in the final result. We define two variables ξ and η, as follows:

$$\xi = 1+\left[1+\frac{R^2+a^2}{L^2}+\frac{2aR}{L^2}\left(1-\frac{y^2}{R^2}\right)^{\frac{1}{2}}\right]^{\frac{1}{2}}, \quad (6.14)$$

$$\eta = 1+\left[1+\frac{R^2+a^2}{L^2}-\frac{2aR}{L^2}\left(1-\frac{y^2}{R^2}\right)^{\frac{1}{2}}\right]^{\frac{1}{2}}. \quad (6.15)$$

With those transformations in eqn (6.12), the integral I_1 becomes

$$I_1 = \int_{\xi_2}^{\xi_1}-\frac{[B^2-A^2+2A(\xi-1)^2-(\xi-1)^4]^{\frac{1}{2}}}{\xi}d\xi$$
$$-\int_{\eta_1}^{\eta_2}\frac{[(B^2-A^2+2A(\eta-1)^2-(\eta-1)^4]^{\frac{1}{2}}}{\eta}d\eta, \quad (6.16)$$

where

$$A = 1+\frac{R^2+a^2}{L^2}, \quad B = 2\frac{Ra}{L^2},$$

and

$$\xi_1 = 1+(A+B)^{\frac{1}{2}}, \quad \xi_2 = 1+A^{\frac{1}{2}},$$
$$\eta_1 = 1-(A-B)^{\frac{1}{2}}, \quad \eta_2 = 1+A^{\frac{1}{2}}.$$

The two components of I_1 have the same integrand but different limits, and by a coincidence the lower limit of the first integral (ξ_2) is the same as the upper limit of the second one (η_2). Therefore the two variables ξ and η may be treated as dummy variables and the two integrals can be treated as one single integral with a new variable t which is equal to ξ in one part of the integral and to η in the other part. The limits of the integration are

$$t_1 = 1 - (A - B)^{\frac{1}{2}}$$

and

$$t_2 = 1 + (A + B)^{\frac{1}{2}},$$

so that

$$I_1 = \int_{1-(A-B)^{\frac{1}{2}}}^{1+(A+B)^{\frac{1}{2}}} -\frac{[B^2 - A^2 + 2A(t-1)^2 - (t-1)^4]^{\frac{1}{2}}}{t} \, dt. \tag{6.17}$$

It is now possible to calculate the integral I_1 as well as I_2. After a lengthy calculation, the result is

$$
\frac{a'_r}{2G\rho} \equiv C_y(L, R, a) = \frac{L^2}{a\left\{\left[1+\left(\frac{R+a}{L}\right)^2\right]^{\frac{1}{2}} + \left[1+\left(\frac{R-a}{L}\right)^2\right]^{\frac{1}{2}}\right\}}
$$

$$
\times \left[\left\{i + 2\frac{R^2+a^2}{L^2} + \left[1+\left(\frac{R-a}{L}\right)^2\right]^{\frac{1}{2}}\right.\right.
$$

$$
\times \left\{\left[1+\left(\frac{R+a}{L}\right)^2\right]^{\frac{1}{2}} + \frac{R^2+a^2}{L^2}\right\}
$$

$$
+ \left(\frac{R^2-a^2}{L^2}\right)^2 \frac{1}{1-\left[1+\left(\frac{R-a}{L}\right)^2\right]^{\frac{1}{2}}}\right\} K(k)
$$

$$
- \frac{1}{2}\left\{\left[1+\left(\frac{R+a}{L}\right)^2\right]^{\frac{1}{2}} + \left[1+\left(\frac{R-a}{L}\right)^2\right]^{\frac{1}{2}}\right\}^2 E(k)
$$

$$
- 2\frac{R^2+a^2}{L^2}\left[1+\left(\frac{R-a}{L}\right)^2\right]^{\frac{1}{2}} \Pi(\tfrac{1}{2}\pi, k, k)
$$

$$
+ 2\frac{(R+a)^2}{L^2}\left[1+\left(\frac{R-a}{L}\right)^2\right]^{\frac{1}{2}} \Pi(\tfrac{1}{2}\pi, \alpha^2, k)\right] + I_0, \tag{6.18}
$$

where

$$I_0 = \begin{cases} \dfrac{\pi R^2}{2a} & \text{when} \quad a \geqslant R, \\[2ex] \dfrac{\pi a}{2} & \text{when} \quad a \leqslant R, \end{cases} \tag{6.19}$$

and

$$k = \frac{\left[1 + \left(\dfrac{R+a}{L}\right)^2\right]^{\frac{1}{2}} - \left[1 + \left(\dfrac{R-a}{L}\right)^2\right]^{\frac{1}{2}}}{\left[1 + \left(\dfrac{R+a}{L}\right)^2\right]^{\frac{1}{2}} + \left[1 + \left(\dfrac{R-a}{L}\right)^2\right]^{\frac{1}{2}}}, \tag{6.20}$$

$$\alpha^2 = \frac{1 - \left[1 + \left(\dfrac{R-a}{L}\right)^2\right]^{\frac{1}{2}}}{1 + \left[1 + \left(\dfrac{R-a}{L}\right)^2\right]^{\frac{1}{2}}} k. \tag{6.21}$$

$K(k)$, $E(k)$ and $\Pi(\frac{1}{2}\pi, k, k)$, $\Pi(\frac{1}{2}\pi, \alpha^2, k)$ here are the complete elliptical integrals of the first, second and third kind respectively (see Abramowitz & Stegun, 1964) and the function $C_y(L, R, a)$ is defined by eqn (6.18).

We have obtained the solution for a point P' which is located in the basal plane of the cylinder. To obtain a solution for an arbitrary point P, we add the attraction of a second cylinder upon the same basal plane. There are three cases.

The first situation is as shown in Fig. 6.4 where P is located in a plane which is within the complete cylinder. In this case, the point $P(a, l)$ may be considered to lie in the basal planes of two cylinders, one of length l and one of length $(L - l)$, and the radial gravitational attraction of the cylinder is given by

$$a_r/2G\rho = C_y(l, R, a) + C_y(L - l, R, a). \tag{6.22a}$$

The second situation is that the point P lies below the basal plane of the cylinder. In this case, the point $P(a, l)$ may be considered to be in the basal planes of a cylinder of positive mass, and length $(L + l)$ and one of negative mass and length l. Then, the attraction is

$$a_r/2G\rho = C_y(L + l, R, a) - C_y(l, R, a). \tag{6.22b}$$

Finally the point P lies above the upper plane of the cylinder, so that $l > L$. Now the attraction is

$$a_r/2G\rho = C_y(l, R, a) - C_y(l - L, R, a). \tag{6.22c}$$

The optimum shape of the cylinder

Because a cylinder is defined by the two parameters R and L, there must be a particular pair of R and L that gives the maximum gravitational efficiency among all cylinders having the same mass. This combination of R and L defines the optimum shape of a cylinder. (For optimum shape of gravitational attracting masses, see also Metherell *et al.* (1984).)

We now discuss the optimum shape of a cylinder for radial gravitational attraction. First we note that such a shape should be dependent upon the value of a. However, to obtain the greatest attraction, the distance of a test mass to an attracting mass should be made as short as possible and therefore we are only interested in the case $a \approx R$.

It may be proved that, when $a \geqslant R$, $C_y(L, R, a)$ has no stationary points with respect to a. Thus we choose the point $a = R$ as the point at which to compare the magnitude of force for all the cylinders, because it is the point of maximum field for every cylinder.

If $a = R$, eqn (6.18) simplifies to

$$a'_r = 2G\rho R\left[\frac{1-k^2}{k}K(k) - \frac{1-k}{k}E(k)\right]. \tag{6.23}$$

To find the ratio R/L, for which the expression (6.23) attains its maximum value under the condition

$$R^2 L = \text{constant} = Q, \tag{6.24}$$

we use the Lagrangian method of undetermined multipliers and find the equation determining the stationary point of (6.23) subject to (6.24) to be

$$\frac{E(k)-K(k)}{k} + kK(k) - E(k)$$

$$+ 48\frac{R^6[K(k)-E(k)]}{Q^2[1+(1+4R^6/Q^2)^{\frac{1}{2}}][1-(1+4R^6/Q^2)^{\frac{1}{2}}]} = 0. \tag{6.25}$$

The parameter k (eqn (6.20)) is

$$k = \frac{(1+4R^6/Q^2)^{\frac{1}{2}}-1}{(1+4R^6/Q^2)^{\frac{1}{2}}+1}. \tag{6.26}$$

and the solution of eqn (6.25) having physical significance is

$$R = 1.009\,667Q^{\frac{1}{3}}, \tag{6.27}$$

or according to eqn (6.24)

$$R = 1.029\,282L. \tag{6.28}$$

It is not difficult to show that this solution gives the maximum value of a'_r.

The superposition principle of gravitation easily shows that if we put the test mass at the middle plane of the cylinder, the optimum form is given by

$$D/L = 1.029282,$$

where D, L are the diameter and length of the cylinder.

Further calculation shows that a cylinder of this shape produces a gravitational attraction close to the attraction produced by a sphere with the same mass and density. Therefore, in practice, we should choose a cylinder with a diameter to length ratio of roughly 1.

As a practical example consider the attraction at a point located in the mid-plane of a cylinder of which the dimensions are:

$$L = 100 \text{ mm},$$
$$D = 2R = 100 \text{ mm},$$

and let the field point be at $a = 75$ mm from the axis of the cylinder. Then the radial gravitational attraction, according to eqn (6.22a), is

$$a_r = (G\rho) \times 125.846 \text{ mm}.$$

However, a sphere having the same mass, density and separation from the field point will produce an acceleration

$$a_s = (G\rho) \times 139.626 \text{ mm}.$$

Thus, the gravitational efficiency, if expressed in the form of the ratio of two attractions, is 90 per cent. That is quite acceptable considering that the cylinder is much easier to manufacture than the sphere. There is a further advantage in using the right cylinder as is now explained.

Relaxation of positional accuracy required with a right circular cylinder

A study of the function $C_y(L, R, a)$ (eqn (6.18)) shows how it is possible to locate a test mass relative to an attracting cylinder so as to alleviate problems of measuring their relative positions. We shall return a number of times to this general problem of easing the task of metrology, but if an experiment is designed to avoid precise measurements of distance, it will usually be at the expense of a smaller attraction, or more demanding requirement in some other respect. The trade off between metrology requirements and magnitude of attraction is often a crucial issue in the design of an experiment and one of the most important considerations for an experimenter is to decide the price worth paying. In this section, we

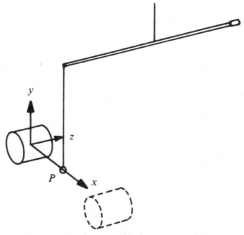

Fig. 6.5 Reduction of precision required in measurement of positions relative to a cylinder.

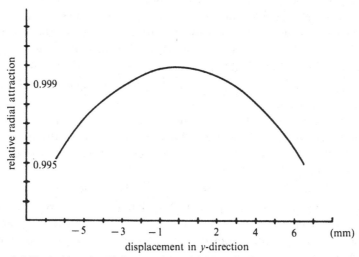

Fig. 6.6 Variation of radial attraction with displacement in the *y*-direction.

discuss a method which can reduce the metrological requirement for distance measurement where it is rather difficult to have very high precision; at the same time it will increase the requirement in other distance measurements which, however, can be done easily with some special method.

Fig. 6.5 shows a test mass suspended from a balance to measure the radial gravitational force produced by a solid cylinder with a length to

Table 6.1. *Difference of radial attraction of a cylinder at a position slightly displaced from y = 0*

Δ (cm)	a_r $(G\rho\text{ m s}^{-2})$	Error (%)
0.00	12.5846	0.000
0.005	12.5842	−0.003
0.10	12.5829	−0.013
0.15	12.5809	−0.029
0.20	12.5782	−0.051

diameter ratio of 1. The positional measurement accuracy required in the experiment can be analysed along three orthogonal directions: x (radial), y (tangential) and z (axial) respectively. The attraction in the x-direction (radial) at the point $(x_0, 0, 0)$ is stationary (maximum) with respect to the small movements in the y- and z-directions. The study of the behaviour of the gravitational field around this stationary point shows that the radial gravitational field varies less than that of a sphere along the x- and y-directions, particularly along the y-direction in which a high precision of positional measurement is difficult to achieve. Fig. 6.6, calculated from eqn (6.22a), shows the behaviour of the radial attraction against the displacement of a test mass along the direction of y.

To illustrate this point, a set of numerical data is given in Table 6.1 for a cylinder of length $L = 100$ mm and radius $R = 50$ mm, with a separation from the test mass of $a = 75$ mm; Δ is the deviation from the attraction at the stationary point $y = 0$.

Thus even at a point departing 1 mm from the mid-level of the cylinder, the attraction is still within 1 part in 10^4 of that at the mid-level. The radial gravitational field a_r varies more along the z-direction than along the y-direction, but it does attain its maximum in the median plane of the cylinder.

The relaxation of the requirement of the position in the y- and z-directions, according to the calculation based on eqn (6.22a), is balanced by a more severe requirement on the measurement of position in the x (radial)-direction. That raises the general question of measuring a distance to an object (a test mass) which is not fixed relative to the cylinder, but is on a mobile balance. It is a point which comes up in most experiments on the inverse square law and the constant of gravitation, and one solution to it is as follows.

Because the x-direction is the direction of the movement of the torsion balance, the separation along such a direction can be measured by using the direct deflection of the torsion balance itself. A second identical cylinder is placed in a position such that the test mass is nearly midway between the two cylinders (as shown in Fig. 6.5). The deviation of the test mass from the mid-point P is indicated by δ. If δ is small compared to the separation of the two cylinders, the study of the function of $C_y(L, R, a)$ shows that δ is proportional to the torque measured by the torsion balance. That means that in the central region between the two cylinders, the gravitational field varies linearly with the displacement and that enables the position P to be found from torque measurements instead of by direct distance measurement. A practical way is to move the two cylinders together relative to the test mass until the torque is the same as when the two cylinders are completely removed. Then the distance x_0 between the test mass and either cylinder is half the separation of the two cylinders and that can be measured with high precision if they are fixed relative to each other.

Radial gravitational attraction inside a hollow cylinder

To complete the discussion of the radial attraction of finite cylinders, we consider the radial attraction inside a hollow cylinder; the formulae of eqn (6.22) are in fact valid for all points including those inside the cylinder.

Consider the radial attraction of a point P inside a hollow cylinder of length L and internal and external radii r_1, r_2. An arbitrary point P is located in the position shown in Fig. 6.7, the vertical distance to the face of the cylinder is l, the radial attraction is given by

$$a_r = 2G\rho[C_y(L-l, r_2, a) + C_y(l, r_2, a) - C_y(L-l, r_1, a) - C_y(l, r_1, a)].$$
(6.29)

This general formula may be used for any hollow cylinder. As an example, consider a long hollow cylinder, i.e. $r_2/L \ll 1$, and two special cases.

First, $l = \frac{1}{2}L$, i.e. the test point P is located in the central plane; the radial acceleration is then (see Chen, 1982)

$$a_r = -\frac{4GM}{L^2}\left(\frac{a}{L} - 3\frac{a^3}{L^3} - 3\frac{r_2^2 + r_1^2}{L^3}a + \ldots\right),$$
(6.30)

where M is the mass of the hollow cylinder. Note that the attraction tends to vanish as a/L goes to zero, the field inside a complete shell is zero everywhere, and not just on the axes ($a = 0$).

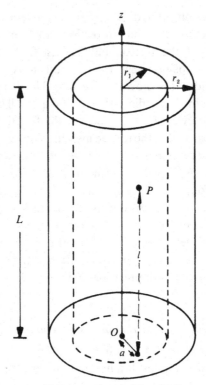

Fig. 6.7 Radial attraction inside a hollow cylinder.

Secondly, it is not easy in practice to place the test mass exactly in the central plane, but, according to the superposition of gravitation, this effect can be calculated with the following formula, provided the test mass is located at a point Q at a small distance Δ from the central plane,

$$a_r = -\frac{1}{2}\frac{GM}{L(\frac{1}{2}L-\Delta)}\left[\frac{a}{\frac{1}{2}L-\Delta}-\frac{3}{4}\frac{a^3}{(\frac{1}{2}L-\Delta)^3}-\frac{3\,a(r_2^2+r_1^2)}{4\,(\frac{1}{2}L-\Delta)^3}-\cdots\right]$$
$$-\frac{1}{2}\frac{GM}{L(\frac{1}{2}L+\Delta)}\left[\frac{a}{\frac{1}{2}L+\Delta}-\frac{3}{4}\frac{a^3}{(\frac{1}{2}L+\Delta)^3}-\frac{3\,a(r_2^2+r_1^2)}{4\,(\frac{1}{2}L+\Delta)^3}-\cdots\right]. \quad (6.31)$$

These formulae are quite useful in practical experiments and will find use in the later part of this book.

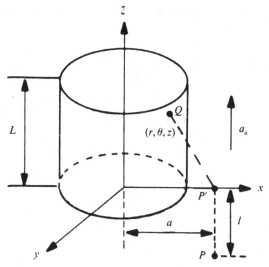

Fig. 6.8 Axial gravitational attraction of a right circular cylinder.

Axial gravitational attraction of a finite cylinder

$C_t(L, R, a)$ function

As in the case of radial attraction, a point P' located in the basal plane of the cylinder is considered and the distance from the axis of the cylinder is again taken to be a (Fig. 6.8); Cartesian coordinates are chosen in such a way that arbitrary points P and P' are located in the (x, z)-plane, as before, this choice of coordinates does not affect the general applicability of our discussion. The length, radius and (constant) density of the cylinder are L, R and ρ respectively; Q is the point of a volume element inside the cylinder, its coordinates in the cylindrical coordinate system being r, θ, z. The gravitational attraction of the cylinder at point P' in the axial direction will be

$$a'_a = \frac{\partial}{\partial z}\left[\int_0^R r\,dr \int_0^\pi d\theta \int_0^L dz \frac{G\rho}{(z^2 + r^2 + a^2 - 2ar\cos\theta)^{\frac{1}{2}}}\right]. \quad (6.32)$$

It is easy to see from the above equation that the triple integral can become two double integrals as follows:

$$a'_a = -2G\rho\left[\int_0^R r\,dr \int_0^\pi \frac{d\theta}{(L^2 + r^2 + a^2 - 2ar\cos\theta)^{\frac{1}{2}}}\right]$$

$$+ 2G\rho\left[\int_0^R r\,dr \int_0^\pi \frac{d\theta}{(r^2 + a^2 - 2ar\cos\theta)^{\frac{1}{2}}}\right]. \quad (6.33)$$

The calculation of the above integrals is very lengthy, but straightforward, and the final result is

$$\frac{a'_a}{2G\rho} \equiv C_t(L, R, a) = -\frac{R^2 - L^2 - a^2}{[(R+a)^2 + L^2]^{\frac{1}{2}}} K(k) - [(R+a)^2 + L^2]^{\frac{1}{2}} E(k)$$

$$- \frac{L^2}{[(R+a)^2 + L^2]^{\frac{1}{2}}} \left[\frac{(L^2 + a^2)^{\frac{1}{2}} + R}{(L^2 + a^2)^{\frac{1}{2}} - a} \Pi(\tfrac{1}{2}\pi, \rho_1, k) \right.$$

$$\left. + \frac{(L^2 + a^2)^{\frac{1}{2}} - R}{(L^2 + a^2)^{\frac{1}{2}} + a} \Pi(\tfrac{1}{2}\pi, \rho_2, k) \right] + L\pi, \tag{6.34}$$

where

$$k = \left[\frac{4aR}{(a+R)^2 + L^2} \right]^{\frac{1}{2}},$$

$$\rho_1 = \frac{-2a}{(L^2 + a^2)^{\frac{1}{2}} - a},$$

$$\rho_2 = \frac{2a}{(L^2 + a^2)^{\frac{1}{2}} + a},$$

and the function $C_t(L, R, a)$ is defined by eqn (6.34).

To generalize the above result to an arbitrary point P, again, the principle of superposition can be used. As with the radial attraction, there are three cases.

The first case is as shown in Fig. 6.8, where P is located in a plane which cuts the cylinder. The point $P(a, l)$ may be considered to lie in the basal planes of a cylinder of length l and a cylinder of length $(L-l)$ and the radial gravitational attraction of the cylinder is

$$a_r/2G\rho = C_t(l, R, a) - C_t(L-l, R, a). \tag{6.35a}$$

Notice the minus sign between the two terms; the axial attractions of these two cylinders are in opposite directions. If $l = L/2$, the resultant attraction is zero.

In the second case the point P is in a plane which lies below the basal plane of the cylinder and so $l < 0$. The point $P(a, l)$ may be considered to lie in the basal planes of a cylinder of length $(L+|l|)$ and one of length $|l|$ and the result is

$$a_r/2G\rho = C_t(L+|l|, R, a) - C_t(|l|, R, a). \tag{6.35b}$$

In the third case, the point $P(a, l)$ is located in a plane which lies above the upper plane of the cylinder and $l > L$; the result is

$$a_r/2G\rho = C_t(l, R, a) - C_t(l-L, R, a). \tag{6.35c}$$

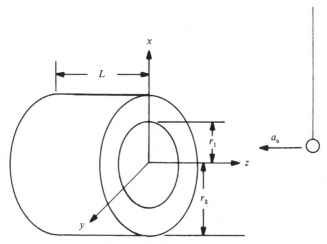

Fig. 6.9 Hollow cylinder.

Behaviour around stationary point of a hollow cylinder

The formula of eqns (6.35a, b, c) can be applied by superposition to a hollow cylinder, thus allowing the behaviour of the attraction around the stationary point of a hollow cylinder to be discussed.

The axial gravitational attraction produced by a hollow cylinder at a point (x, y, z), as shown in Fig. 6.9 is

$$a_a = 2G\rho\{C_t[z+L, r_2, (x^2+y^2)^{\frac{1}{2}}] - C_t[z, r_2, (x^2+y^2)^{\frac{1}{2}}]$$
$$- C_t[z+L, r_1, (x^2+y^2)^{\frac{1}{2}}] + C_t[z, r_1, (x^2+y^2)^{\frac{1}{2}}]\}, \quad (6.36)$$

where L, r_1 and r_2 are respectively the length, internal radius and external radius of the hollow cylinder. The stationary points of the axial attraction are determined by the conditions

$$\frac{\partial a_a}{\partial x} = 0, \tag{6.37a}$$

$$\frac{\partial a_a}{\partial y} = 0, \tag{6.37b}$$

$$\frac{\partial a_a}{\partial z} = 0, \tag{6.37c}$$

and, by symmetry, those points must lie on the axis:

$$x = 0,$$

$$y = 0.$$

The previous calculation of the attraction at a point on the axis shows that the attraction of a hollow cylinder on the axis is proportional to

$$-A+B+C-D,$$

where

$$A = [(z+L)^2+r_2^2]^{\frac{1}{2}}, \tag{6.38a}$$

$$B = (z^2+r_2^2)^{\frac{1}{2}}, \tag{6.38b}$$

$$C = [(z+L)^2+r_1^2]^{\frac{1}{2}}, \tag{6.38c}$$

$$D = (z^2+r_1^2)^{\frac{1}{2}}, \tag{6.38d}$$

and so the position of the stationary point is determined by

$$-\frac{\partial A}{\partial z}+\frac{\partial B}{\partial z}+\frac{\partial C}{\partial z}-\frac{\partial D}{\partial z}=0, \tag{6.39}$$

that is,

$$\frac{z_0+L}{[(z_0+L)^2+r_2^2]^{\frac{1}{2}}}-\frac{z_0}{[z_0^2+r_2^2]^{\frac{1}{2}}}-\frac{z_0+L}{[(z_0+L)^2+r_1^2]^{\frac{1}{2}}}+\frac{z_0}{[z_0^2+r_1^2]^{\frac{1}{2}}}=0. \tag{6.40}$$

An idea of the behaviour of the axial gravitational attraction in the (x,y,z) space around the stationary point $(0,0,z_0)$ may be obtained by expanding the expression for a_a (eqn (6.36)), omitting terms of higher order than y^2, x^2 and $(z-z_0)^2$ to give

$$a_a = 2G\rho\{F+G[x^2+y^2-2(z-z_0)^2]\}, \tag{6.41}$$

where F and G are the constants given by

$$F = -A_0+B_0+C_0-D_0 \tag{6.42a}$$

and

$$G = \frac{1}{4}\left(\frac{r_2^2}{A_0^3}-\frac{r_2^2}{B_0^3}-\frac{r_1^2}{C_0^3}+\frac{r_1^2}{D_0^3}\right), \tag{6.42b}$$

and A_0, B_0, C_0 and D_0 are the values of A, B, C, D (eqns (6.38a–d)) with z replaced by z_0.

Eqn (6.41) shows that the behaviour of the gravitational field of a_a around the stationary point is a hyper-saddle surface in a four-dimensional (a_a, x, y, z) space, but because of the symmetry with respect to the two coordinates x and y, it can equally be represented by a surface in the three dimensions of a_a, z and radial distance x (Fig. 6.10). The stationary point $(0,0,z_0)$ is the point at which the gravitational field takes its maximum value along the z-axis, and its minimum value along the radial directions (y-axis or x-axis) (Chen & Cook, 1989).

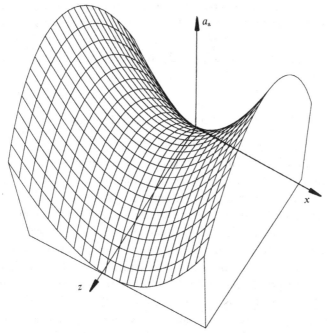

Fig. 6.10 Attraction of thick ring near a saddle point.

That above behaviour is very important in practice, because the measurement of the precise distance between an attracting mass and an attracted mass is difficult as already mentioned, especially when the attracted mass is inside a vacuum chamber. The uncertainty of that measurement may sometimes dominate the final result. The use of a hollow cylinder avoids the problem to a large extent, since the mathematical behaviour of the gravitational field around the stationary point allows us to determine the position of the attracting mass from the force on the test mass itself and so from the deflection of the torsion balance. To do that, the dimensions of the hollow cylinder must be very accurately measured and the stationary point $(0,0,z_0)$ must be very precisely computed using eqn (6.40). The calculated value z_0 is the distance at which the test mass should be placed.

The attracting hollow cylinder is then moved relative to the attracted mass (suspended, for example, from a torsion balance) along the x, y and z directions to find the right position. When it moves in the x- and y-directions, the minimum attraction indicates that the test mass is located at the point $(0,0)$; when it moves in the z-direction, the maximum attraction indicates that the test mass is located at the point z_0.

It must be emphasized here that the cylinder must be adjusted along three axes to find the right point. In some previous experiments, attention was paid only to the maximum value along the z-axis and the minimum along the x- and y-axes was ignored; that may have caused errors in the measurement.

6.4 Attraction of a rectangular prism

Introduction

It is sometimes convenient to use rectangular prismatic blocks to develop a gravitational field in the laboratory for not only are blocks of that form easy to manufacture, but also it is easy to pack many identical blocks together to make a large mass to produce a strong gravitational field. It is difficult to make a single sphere or a cylinder precisely and, because of their curved surfaces, they cannot be packed tightly together. In addition, because the sides of a prism are flat, optical interferometric techniques may be used to measure the dimensions and separations of rectangular blocks.

MacMillan (1958) has already summarized the formulae for the attraction of a two-dimensional uniform rectangular plant on a point in its own plane, and for the gravitational potential of a three-dimensional homogeneous rectangular parallelepiped at an arbitrary point. However, as the formulae are relatively complicated, it may be more convenient to have an explicit set of formulae for the gravitational attractions in the three orthogonal directions produced by a three-dimensional uniform rectangular block.

As shown in Fig. 6.11, a Cartesian coordinate system (x, y, z) is taken with its origin in the centre of a homogeneous rectangular parallelepiped and with the axes parallel to the three sides. The lengths of the three sides are $2a$, $2b$ and $2c$ respectively. The three components of the gravitational attraction produced by the block at an arbitrary point $P(x, y, z)$ are given by

$$X = -G\rho \frac{\partial}{\partial x} \left\{ \int_{-a}^{a} \int_{-b}^{b} \int_{-c}^{c} \frac{dw\,dv\,du}{[(x-u)^2+(y-v)^2+(z-w)^2]^{\frac{1}{2}}} \right\}, \quad (6.43\,a)$$

$$Y = -G\rho \frac{\partial}{\partial y} \left\{ \int_{-a}^{a} \int_{-b}^{b} \int_{-c}^{c} \frac{dw\,dv\,du}{[(x-u)^2+(y-v)^2+(z-w)^2]^{\frac{1}{2}}} \right\}, \quad (6.43\,b)$$

$$Z = -G\rho \frac{\partial}{\partial z} \left\{ \int_{-a}^{a} \int_{-b}^{b} \int_{-c}^{c} \frac{dw\,dv\,du}{[(x-u)^2+(y-v)^2+(z-w)^2]^{\frac{1}{2}}} \right\}, \quad (6.43\,c)$$

where ρ is the density (which is assumed to be a constant) and u, v, w are

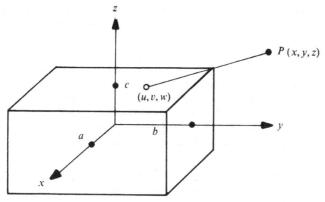

Fig. 6.11 Rectangular parallelepiped.

the coordinates of an element of volume of the block. The integrands in the above equations all have the same form, and the integrations are similar. Eqn (6.43a) for the attraction in the x-direction is considered in detail.

After differentiating with respect to x and integrating with respect to u, eqn (6.43a) becomes

$$X = G\rho \left\{ \int_{-b}^{b} \int_{-c}^{c} \frac{dw\,dy}{[(x-a)^2+(y-v)^2+(z-w)^2]^{\frac{1}{2}}} \right.$$

$$\left. - \int_{-b}^{b} \int_{-c}^{c} \frac{dw\,dy}{[(x+a)^2+(y-v)^2+(z-w)^2]^{\frac{1}{2}}} \right\}$$

$$\equiv G\rho[M(x-a,y-b,y+b,z-c,z+c)$$

$$- M(x+a,y-b,y+b,z-c,z+c)], \qquad (6.44)$$

where $M(x-a,y-b,y+b,z-c,z+c)$, the integral with respect to u, is a function which depends on the dimensions of the rectangular block and the position of the test mass.

The function $M(x-a,y-b,y+b,z-c,z+c)$

We may separate the function M of eqn (6.44) into three parts as

$$M = \int_{-b}^{b} \int_{-c}^{c} \frac{(x-a)^2+(y-v)^2}{[(x-a)^2+(y-v)^2+(z-w)^2]^{\frac{3}{2}}} dw\,dv$$

$$+ \int_{-b}^{b} \int_{-c}^{c} \frac{(x-a)^2+(z-w)^2}{[(x-a)^2+(y-v)^2+(z-w)^2]^{\frac{3}{2}}} dw\,dv$$

$$- \int_{-b}^{b} \int_{-c}^{c} \frac{(x-a)^2}{[(x-a)^2+(y-v)^2+(z-w)^2]^{\frac{3}{2}}} dw\,dv. \qquad (6.45)$$

Following the method given by MacMillan (1958), the final form of the function M is

$$M(x-a, y-b, y+b, z-c, z+c)$$

$$= (y-b)\ln\frac{z-c+[(x-a)^2+(y-b)^2+(z-c)^2]^{\frac{1}{2}}}{z+c+[(x-a)^2+(y-b)^2+(z-c)^2]^{\frac{1}{2}}}$$

$$-(y+b)\ln\frac{z-c+[(x-a)^2+(y+b)^2+(z-c)^2]^{\frac{1}{2}}}{z+c+[(x-a)^2+(y+b)^2+(z-c)^2]^{\frac{1}{2}}}$$

$$+(z-c)\ln\frac{y-b+[(x-a)^2+(y-b)^2+(z-c)^2]^{\frac{1}{2}}}{y+b+[(x-a)^2+(y-b)^2+(z-c)^2]^{\frac{1}{2}}}$$

$$-(z+c)\ln\frac{y-b+[(x-a)^2+(y-b)^2+(z+c)^2]^{\frac{1}{2}}}{y+b+[(x-a)^2+(y-b)^2+(z+c)^2]^{\frac{1}{2}}}$$

$$-(x-a)\tan^{-1}\frac{(y-b)(z-c)}{(x-a)[(x-a)^2+(y-b)^2+(z-c)^2]^{\frac{1}{2}}}$$

$$+(x-a)\tan^{-1}\frac{(y+b)(z-c)}{(x-a)[(x-a)^2+(y-b)^2+(z-c)^2]^{\frac{1}{2}}}$$

$$+(x-a)\tan^{-1}\frac{(y-b)(z+c)}{(x-a)[(x-a)^2+(y-b)^2+(z-c)^2]^{\frac{1}{2}}}$$

$$-(x-a)\tan^{-1}\frac{(y+b)(z+c)}{(x-a)[(x-a)^2+(y-b)^2+(z-c)^2]^{\frac{1}{2}}}. \qquad (6.46)$$

Components of attraction

The three components of the attraction of a rectangular block are given in terms of the function M as follows:

$$X = G\rho[M(x-a, y-b, y+b, z-c, z+c)$$
$$- M(x+a, y-b, y+b, z-c, z+c)], \quad (6.47a)$$

$$Y = G\rho[M(y-b, z-c, z+c, x-a, x+a)$$
$$- M(y+b, z-c, z+c, x-a, x+a)], \quad (6.47b)$$

$$Z = G\rho[M(z-c, x-a, x+a, y-b, y+b)$$
$$- M(z+c, x-a, x+a, y-b, y+b)]. \quad (6.47c)$$

Three forms of attracting mass, sphere, cylinder and rectangular prism, have now been discussed, but the choice of attracting masses in gravitational experiments is not restricted to them. Various combinations may

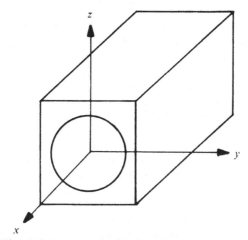

Fig. 6.12 Rectangular block with cylindrical hole.

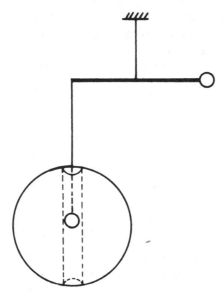

Fig. 6.13 Sphere with hole for access to test mass.

be useful, for example, a rectangular block with a cylindrical hole in the middle, as shown in Fig. 6.12, may produce suitable stationary characteristics; a mass formed by a sphere with a narrow hole in it, as shown in Fig. 6.13, may provide easy access for a test body. The resultant gravitational force in such combinations may be obtained by superposition using the appropriate formulae.

7

Experimental tests of the inverse square law

7.1 Theoretical background

According to legend the fall of an apple prompted Isaac Newton to propose the inverse square law of gravitation. He later wrote: 'there is a power of gravity pertaining to all bodies, proportional to the several quantities of matter which they contain. The force of gravity towards the several equal parts of any body is inversely as the square of the distance of places from the particles' (1687). Accordingly, the force of attraction between two point particles of masses M and m separated by a distance r is

$$F = G\frac{Mm}{r^2}$$ (7.1)

and the gravitational potential is

$$V = -G\frac{M}{r}.$$ (7.2)

The motions of celestial bodies show that the gravitational force obeys the inverse square law very closely at great distances. It is also to a high degree independent of the material nature of bodies as discussed in Chapters 4 and 5. In the neighbourhood of very massive bodies the inverse square law is modified in accordance with general relativity, as shown by the motions of the terrestrial planets about the Sun, but so far no deviation from the inverse square law has been established in laboratory experiments. Experimental tests of the law are the subject of this chapter.

General relativity is a geometrical theory and it is supposed that particles move along geodesic paths in the four dimensions of space and time. Once the metric of the four-dimensional space is known, the motion of the particle is determined by eqn (4.4). The metric is determined by the energy in a region of space as expressed in the field equation

$$R_{\mu\nu} - \tfrac{1}{2} g_{\mu\nu} R = -\frac{8\pi G}{c^4} T_{\mu\nu}, \qquad (7.3)$$

where $R_{\mu\nu}$ is the Ricci tensor of the metric, $g_{\mu\nu}$ is the metric tensor, R is the curvature scalar and $T_{\mu\nu}$ is the energy–momentum tensor.

It is well known that the metric in the neighbourhood of an isolated spherical mass M is 'the Schwarzschild metric'

$$ds^2 = \left(1 - 2\frac{GM}{rc^2}\right)(dx^0)^2 - \frac{dr^2}{\left(1 - 2\dfrac{GM}{rc^2}\right)} - r^2 d\theta^2 - r^2 \sin^2\theta \, d\phi^2. \quad (7.4)$$

Here ds is the space-time interval, x^0 is ct, θ and ϕ are the colatitude and azimuthal angles and c is the speed of light.

In a weak field, GM/rc^2 is small, g_{00} may be taken to be $(1 - 2GM/rc^2)$ and the equation of motion (eqn (4.4)) may be shown to reduce to

$$\frac{d^2 x_i}{dt^2} = -\delta_{ij} \frac{\partial [c^2(g_{00} - 1)]}{\partial x_j}. \qquad (7.5)$$

That is just Newton's equation for motion in a potential given by

$$V = \tfrac{1}{2} c^2 (g_{00} - 1), \qquad (7.6)$$

and with the Schwarzschild value of g_{00},

$$V = -G\frac{M}{r}, \qquad (7.7)$$

so that the inverse square law is the first approximation with the Schwarzschild metric.

It can readily be seen that in higher-order approximation, the factor $(1 - 2GM/rc^2)^{-1}$ that multiplies dr^2 must be taken into account, and that leads to higher powers of r^{-1} in the potential, with the well-known consequence of a contribution to the motion of the perihelion of Mercury and other planets from the r^{-2} term in the potential. The gravitational inverse square law is thus the direct consequence of general relativity and certainly applicable in the weak fields of laboratory experiments. However, two significant criticisms of general relativity have been made.

First, the theory is supported by rather meagre experimental evidence. The evidence can be summarized as (*a*) the perihelion precession of planets (see Duncombe, 1956), (*b*) the deflection of light and radio waves by the Sun (Eddington, 1920), (*c*) the time delay of radar pulses reflected from the planets (Shapiro *et al.*, 1972a; Shapiro, 1980) and the gravitational red shift of light (see Snider, 1972). A recent summary is that of Will (1987). For this reason, many new parallel theories of gravitation by which the above experiments can be similarly explained have been proposed. Will and Nordvedt (see Will, 1981) developed a general way of expressing deviations from general relativity in Parameterized Post Newtonian (PPN) theories of gravitation (see also Thorne *et al.*, 1973).

Secondly, general relativity has very little connection with the rest of physics (quantum mechanics, elementary particle physics, etc.). A great deal of theoretical effort, so far unsuccessful, has been put into unifying gravity with other forces, namely the weak electromagnetic force and the strong force. On general grounds it is natural to assume that the gravitational force corresponds to an exchange of particles. A potential produced by the exchange of massless particles corresponds to Laplace's equation in free space, that is,

$$\nabla^2 V = 0, \tag{7.8}$$

from which the inverse square law follows. If, however, the exchange particles have mass m then, to be consistent with special relativity, their energies and momentum must satisfy the energy–momentum relation of special relativity,

$$E^2 - c^2 p^2 = E_0^2, \tag{7.9}$$

where E is energy, p is momentum, E_0 is the rest energy.

The equivalent field equation is the Klein–Gordon equation,

$$\nabla^2 \phi - \frac{\partial^2 \phi}{\partial t^2} = \mu^2 \phi \tag{7.10}$$

with

$$\mu = \frac{mc}{h}. \tag{7.11}$$

If the field is stationary and symmetric with respect to colatitude (θ) and azimuth (ϕ), the Klein–Gordon equation reduces to

$$\frac{\partial^2 \phi}{\partial r^2} - \frac{2}{r}\frac{\partial \phi}{\partial r} = \mu^2 \phi, \tag{7.12}$$

which has the solution

$$\phi = -\frac{C_1}{r}e^{-\mu r}, \tag{7.13}$$

where C_1 is a constant. The most general form of the potential energy of two point masses M and m satisfying Lorentz invariance, is then

$$V = -G_0 \frac{Mm}{r}(1 + \alpha e^{-\mu r}), \tag{7.14}$$

where G_0 is the gravitational constant at a great distance; the second term in the bracket is called the Yukawa term with the coupling constant α.

Various authors have proposed a law of that form. Fujii (1971, 1972) from a discussion of scale invariance, one of the main principles in elementary particle physics, obtained the value of $\frac{1}{3}$ for the parameter α. O'Hanlon (1972) derived an equation of the same form from the scalar–tensor theory. Acharya & Hogan (1973) predicted a similar form by introducing a finite mass for the scalar field in the Brans–Dicke theory. Scherk (1979) obtained the same result from the theory of supergravity.

Fischbach *et al.* (1986) have also proposed a similar contribution to the effective gravitational potential, supposing that it was proportional to the product of the baryon numbers of attracting bodies. Now the baryon number is very nearly proportional to atomic mass but not exactly so because of differences in the binding energies of nucleons, and so Fischbach *et al.* anticipated that the weak principle of equivalence would be violated and as a consequence that the inverse square law would not be exactly obeyed.

We may see that in designing experimental tests, it is essential to distinguish the violation of the inverse square law discussed in this chapter from the relativistic effect and from the so-called 'fifth force' effect originating from the theory of Fischbach *et al.* The violation of the inverse square law is directly related to the law itself regardless of the material used. The 'fifth force' effect refers to the use of attracting masses made of different materials, the violation of the inverse square law is a consequence. The discussion of this chapter is, however, restricted to experimental tests of the inverse square law in which the test bodies are made of the same material (or if different materials are used, any difference between materials is ignored). In this sense, we name the deviation from the inverse square law 'non-Newtonian behaviour'. The purpose of the experiments described in the following is to look for non-Newtonian behaviour on a laboratory scale.

7.2 Interpretation of the experimental results

If the attraction is not inverse square, then the formulae for the potentials and attractions derived in Chapter 6 will not apply and thus non-Newtonian gravitational behaviour may occur both on account of the shapes of the attracting and attracted masses as well as through the separations between them. For this and other reasons that will appear, an accurate interpretation of any experimental test on non-Newtonian gravitation is complex. Two ways of interpreting and comparing the results of different experiments will be used in this book.

Plot method

A widely adopted way of expressing experimental results was introduced by Spero *et al.* (1980) and exploited by Gibbons & Whiting (1981). The exponential law of eqn (7.14) contains two parameters, so that a single experiment in which gravitational forces are compared at two distances can only determine a set of pairs, α and μ^{-1}, which are consistent with the experimental observations. Therefore that set may be conveniently shown as a region in the (α, μ^{-1})-plane. This is called the α–μ^{-1} plot method and is now explained in detail (Fig. 7.1).

From the non-Newtonian potential expressed by eqn (7.14), the effective gravitational constant $G(r)$ at a distance r is

$$G(r) = G_0[1 + \alpha(1 + \mu r)e^{-\mu r}]. \tag{7.15}$$

Let the measured value of the ratio $G(r_1)/G(r_2)$ at distances r_1 and r_2 be denoted by Γ; α and μ^{-1} are then constrained to lie on a curve in the (α, μ^{-1})-plane given by

$$\alpha = \frac{\Gamma - 1}{(1 + \mu r_1)e^{-\mu r_1} - \Gamma(1 + \mu r_2)e^{-\mu r_2}}. \tag{7.16}$$

The Newtonian value of Γ is 1 but there will be a range of other values of Γ consistent with the uncertainties of an experiment. Varying Γ within the uncertainties of the experimental data causes the curve to sweep out an allowed region in the (α, μ^{-1})-plane where the non-Newtonian gravitational

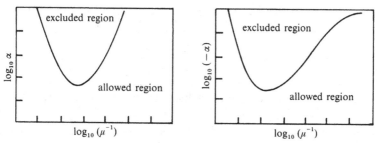

Fig. 7.1 α–μ^{-1} plot for interpretation of experimental results.

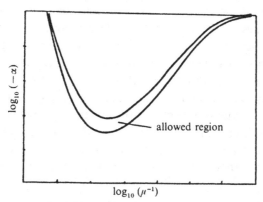

Fig. 7.2 Departure from Newtonian law indicated by closed region on plot.

law may be valid. Outside that region there lie values of α and μ that define an 'excluded' region in the plane consistent with the inverse square law. Thus, a curve can be found that divides the (α, μ^{-1})-plane into two regions: the excluded region and the allowed region (Fig. 7.1). An experimental result which claims to support a breakdown of the inverse square law corresponds to two bounding curves, with the same sign of α and therefore in the same part of the α–μ^{-1} plots; the specific location, though, depends on α being positive or negative, and forming a closed allowed region in the plane as shown in Fig. 7.2.

Analytical method

Experimental results may also be compared analytically. Although the value of the wavelength μ^{-1} of any assumed exchange particle is quite uncertain, yet provided the coupling constant α is not very small (in Fujii's theory (1972), $\alpha = \frac{1}{3}$), it is safe to assume that, under laboratory conditions,

$$\mu r \ll 1. \tag{7.17}$$

Thus, any failure of the inverse square law according to eqn (7.15) gives approximately

$$G(r) = G_0\left(1+\alpha-\frac{\alpha\mu^2r^2}{2}\right) \tag{7.18}$$

and so, if we compare the effective value G at two separations r_1 and r_2,

$$\frac{\Delta G}{G_0} = \frac{G(r_1)-G(r_2)}{G_0} = \frac{\alpha\mu^2}{2}(r_2^2-r_1^2). \tag{7.19}$$

That shows that the relative change of gravitational constant G is proportional to the product of α and μ^2. It must be emphasized that this expression applies to point masses but for any practical mass with an extended volume, the conclusion that the relative change of G is proportional to $\alpha\mu^2$ can be expected to remain correct. That may be seen to be so because if for any two points P and Q located inside the attracting and attracted masses, the non-Newtonian interaction obeys the relation of eqn (7.19) with a certain value of $\alpha\mu^2$, then by the superposition principle, the non-Newtonian interaction for the two bodies with extended volumes, would obey the same relation of eqn (7.19) with the only difference a geometrical factor that will vary with the shapes of the masses, that is,

$$\frac{\Delta G}{G_0} = \alpha\mu^2\tilde{R}^2, \tag{7.20}$$

where $\tilde{R}$ is a characteristic length determined by the shapes and separation of the masses. We give some examples of the specific form of that equation below.

The advantage of using the parameter $\alpha\mu^2$ to compare different experiments is that it takes account both of the experimental precision and of the distance between the attracted mass and attracting mass. If the experimental result is null in searching for a non-Newtonian force, how do we assess the value of the experiment? The final uncertainty of an experiment is not necessarily a guide to the care and imagination that went into it. The experimental precision could be very high when the distance is relatively short (for example, if it is about 0.1 m), and it could be poor when the distance is much greater (for example, if it is several metres). Equation (7.20) shows that the uncertainty in $\alpha\mu^2$ depends both on the uncertainty of $\Delta G/G_0$ and on the characteristic length $\tilde{R}$ which is function of the separation of the masses. The analytical method using $\alpha\mu^2$ can reflect both the experimental precision and the geometrical distance factor.

7.3 Review of experimental studies

As in other kinds of gravitational experiment, the torsion balance is still the main detector used in the laboratory to verify the inverse square law, and of the possible ways in which it may be used (see Chapter 8), direct observations of the deflection due to a steady gravitational force are by far the most common. While noise as considered in Chapters 2 and 3 is important, metrological problems are often more significant in experiments on the inverse square law.

Although Cavendish (1798) stated that he checked the inverse square law in the course of his determination of G, he gave no details, and it seems that no investigations of the law were undertaken in the nineteenth century. Baily (1842, 1843) made no reference to examining the law, nor did Cornu & Baille (1873, 1878) in their respective determinations of the constant of gravitation, nor did any of those responsible for the later measurements referred to in Chapter 8. Some indications of the validity of the law on a laboratory scale can naturally be obtained from the general agreement between determinations of G made with masses at different separations, different mass geometry and different weight of the masses. Thus the results of Braun (1897a, b) and of Boys (1895) agree well, although the dimensions of Boys's apparatus are less than those of Braun's by a factor of about ten. No doubt during the later part of the nineteenth century it was considered that the inverse square law was quite securely established in celestial mechanics and there was no need to check it in the laboratory.

The first reported explicit test of the inverse square law was that of Mackenzie (1895). He concluded, from experiments with a torsion balance on directional effects with crystals, that the inverse square law was valid to within the limit of experimental error then obtainable, which was about 1 per cent at distances of 30 to 70 mm. After this first laboratory test, no further experimental work was done until that of Long (1976) which stimulated a great deal of subsequent experimental work.

Prompted by the positive result of Long (see section 7.6 for details), many groups undertook other experiments. Most supported the inverse square law within their limits of experimental error but some geophysical evidence (Stacey & Tuck, 1981) gave indications of deviations from it at the scale of a few kilometres. Although geophysical measurements can be used to test the inverse square law, the results over short distances in the laboratory are still considered the most reliable evidence. Some results of laboratory measurements are given in Table 7.1.

Table 7.1. *Some results of tests of the inverse square law*

Group	Year	Method	Distance (m)	$\alpha\mu^2$ (m^{-2})	α sign
Washington (Long)	1976	Torsion balance with ring masses	0.0025–0.0175	> 0.07 < 0.1	+ +
Taiwan (Yu et al.)	1979	Gravimeter on oil tank	1–15	< 0.1	+
Moscow (Panov & Frontov)	1979	Torsion balance spherical masses	0.4–3 0.4–10	< 0.001 < 0.0003	+ −
Tokyo (Ogawa et al.)	1982	Resonant plate	2.6–10.7	< 0.03	+
Irvine (Spero et al.)	1980	Torsion balance hollow cylinder	0.02–0.05	< 0.029	+
Irvine (Hoskins et al.)	1985	Torsion balance	0.04–1.15	< 0.0013	+
Cambridge (Chen et al.)	1984	Torsion balance cylinder	0.075–0.17 0.11–0.25	< 0.028 < 0.014	+ +
Beijing (Liu, Zhang & Qin)	1983	Torsion balance	0.04–0.05		+
NBS (USA) (Luther & Towler)	1982	Period of oscillation	0.07–0.1		+
Maryland (Sinsky & Weber)	1967	Resonant bar	1.72–1.84	< 0.1	+
Maryland (Chan et al.)	1982	Paik's method	1.9–2.7	< 0.06	+

A gravitational force on the laboratory scale is so weak that the accuracy of the measurement is poor at separations of little more than 1 m, and the most precise studies have concentrated on distances of about 0.1 m.

It is also important to check the law at ranges of millimetres or less, partly because some theories (e.g. Fujii, 1972) predicted that there might be a violation of the inverse square law in this region, and also because a successful experiment at this very short distance has not yet been published. Such experiments are difficult because at a short distance the influence of surface forces may be considerable, and also because the attracted and attracting masses have to be very small or very thin if the effective separation is to be just a few millimetres; in that event, the attraction will be weak.

7.4 Non-Newtonian fields of extended bodies

It might be thought that tests of the inverse square law in the laboratory could be made quite simply by placing attracting masses at different distances from an attracted mass and comparing the forces with those to be expected from the inverse square law. However, eqn (7.14) applies only to two point masses M and m; in any real laboratory experiment, the extent of the masses will be comparable with their separation and that not only alters the form of the Newtonian attraction, as already seen in Chapter 6, but will also alter the form of the non-Newtonian attraction. The resulting non-Newtonian force can be found only by integrating throughout the whole volumes of the masses. This is very significant in the interpretation of the experiments. An analytical expression for a non-Newtonian force of an extended object is usually very difficult to obtain and numerical methods may be necessary. There are two effects of non-Newtonian forces in laboratory experiments, one from the separation of the attracting masses from the attracted mass, the other from the geometric shapes of the masses. We give here a few examples of analytical forms of a non-Newtonian force, for they are instructive for the design of experiments; for instance, even a perfect sphere cannot be treated as a point mass.

Non-Newtonian attraction of a sphere

Thin spherical shell

Although possibly physically unrealistic, we start with a thin spherical shell because the results for thick shells and solid spheres may be built up from the following formula for thin shells.

(1) *Non-Newtonian attraction outside a spherical shell.* As shown in Fig. 7.3, we may assume quite generally that the test body is located on the z-axis. This choice is only for the convenience of calculation and will not affect the general applicability of the result, as the coordinate system may be chosen in an arbitrary way. With the symbols as in Fig. 7.3, if $a > R$ and $m = 1$ the potential of the non-Newtonian component can be expressed as

$$V_n = -2\pi G_0 \alpha \rho R^2 \int_0^\pi \frac{e^{-\mu x}}{x} \sin \theta \, d\theta, \qquad (7.21)$$

where ρ is the surface density of the shell, R is the radius of the shell, a is

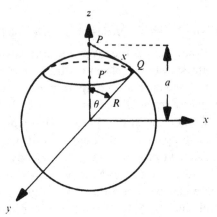

Fig. 7.3 Spherical shell.

the distance of the test mass to the centre of the shell, and x is the distance between the field point P and a point Q in the mass. Now

$$x = (a^2 + R^2 - 2aR\cos\theta)^{\frac{1}{2}},$$

so that

$$V_n = -\frac{2\pi G_0\,\alpha\rho R}{a}\int_{a-R}^{a+R}\mathrm{e}^{-\mu x}\,\mathrm{d}x, \tag{7.22}$$

and the potential of the whole shell is

$$V_n = -\alpha G_0\,M\frac{\sinh\mu R}{\mu R}\frac{\mathrm{e}^{-\mu a}}{a}. \tag{7.23}$$

On differentiating with respect to a, it will be found that the non-Newtonian component of the force is

$$F_n = \alpha G_0\,M\frac{\sinh\mu R}{\mu R}(1+\mu a)\frac{\mathrm{e}^{-\mu a}}{a^2}. \tag{7.24}$$

When μR is much less than 1, the practical situation, $\sinh\mu R/\mu R$ is almost 1. Then the non-Newtonian attraction of the spherical shell is nearly the same as that of a point mass at the centre (see eqn (7.14)). A similar conclusion holds for all spherical attracting masses for a point outside the mass.

(2) *Non-Newtonian attraction inside a spherical shell.* With the same coordinate system as shown in Fig. 7.3, if $a < R$ and $m = 1$, the potential of the non-Newtonian component inside a spherical shell is

$$V_n = -2\pi G_0\,\alpha\rho R\int_{R-a}^{R+a}\frac{\mathrm{e}^{-\mu x}}{a}\,\mathrm{d}x. \tag{7.25}$$

After integration,

$$V_n = \alpha G_0 M \frac{e^{-\mu R}}{R} \frac{\sinh \mu a}{\mu a} \qquad (7.26)$$

and the force acting on the test body is

$$F_n = \alpha G_0 M \frac{e^{-\mu R}}{R} \frac{\mu a \cosh \mu a - \sinh \mu a}{\mu a^2} \qquad (7.27)$$

or approximately

$$\tfrac{1}{3}\alpha G_0 M \frac{e^{-\mu R}}{R} \mu^2 a.$$

There is now a significant difference between the Newtonian law and the non-Newtonian law. The Newtonian gravitational force is null inside a spherical shell while the non-Newtonian component is not null, as shown in eqn (7.27), but to first order proportional to a, the distance from the centre. We shall now see that the same conclusion holds for a sphere with a thick shell.

Thick spherical shell

(1) *Non-Newtonian attraction outside a thick spherical shell.* The non-Newtonian gravitational potential of a thick shell is obtained straightforwardly by integrating the potential for a thin shell between the limits of the inner and outer radii. The notation is as in Fig. 7.4. If r is the radius of an elementary shell, the mass of the shell is $4\pi r^2 \rho \, dr$, and so the non-Newtonian potential from eqn (7.23) is

$$V_n = -4\pi \alpha G_0 \frac{e^{-\mu a}}{\mu a} \int_{R_1}^{R_2} r \sinh \mu r \, dr, \qquad (7.28)$$

where R_1 and R_2 are the internal and external radii respectively; the result is

$$V_n = -\frac{3\alpha G_0 M}{\mu^3(R_2^3 - R_1^3)}$$

$$\times (\mu R_2 \cosh \mu R_2 - \sinh \mu R_2 - \mu R_1 \cosh \mu R_1 + \sinh \mu R_1) \frac{e^{-\mu a}}{a}, \qquad (7.29)$$

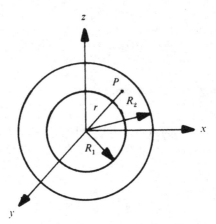

Fig. 7.4 A thick spherical shell.

and so the force is

$$F_n = \frac{3\alpha G_0 M}{\mu^3(R_2^3 - R_1^3)}$$

$$\times (\mu R_2 \cosh \mu R_2 - \sinh \mu R_2 - \mu R_1 \cosh \mu R_1 + \sinh \mu R_1)(1 + \mu a)\frac{e^{-\mu a}}{a^2}.$$

$$(7.30)$$

The function of R_2 and R_1 is 1 to first order, so again we recover the potential and force of a point mass at distance a.

(2) *Non-Newtonian attraction inside a thick spherical shell.* Using the same method, we can obtain the non-Newtonian potential and force at the point inside a thick spherical shell which are given by

$$V_n = -\frac{3\alpha G_0 M}{\mu^3(R_2^3 - R_1^3)}[(1 + \mu R_2)e^{-\mu R_2} - (1 + \mu R_1)e^{-\mu R_1}]\frac{\sinh \mu a}{a} \quad (7.31)$$

and

$$F_n = -\frac{3\alpha G_0 M}{\mu^3(R_2^3 - R_1^3)}[(1 + \mu R_2)e^{-\mu R_2} - (1 + \mu R_1)e^{-\mu R_1}]$$

$$\times \frac{\mu a \cosh \mu a - \sinh \mu a}{a^2}. \quad (7.32)$$

Once again, the force is proportional to a, while the R factor is

$$(R_2 + R_1)/(R_2^2 - R_1 R_2 + R_1^2).$$

(3) *Solid sphere.* The solid sphere is the limiting case of a thick spherical shell and, of course, only the external potential and attraction are relevant. If we let $R_1 = 0$ and $R_2 = R$, then the potential is

$$V_n = -\frac{3\alpha G_0 M}{\mu^3 R^3}(\mu R \cosh \mu R - \sinh \mu R)\frac{e^{-\mu a}}{a} \tag{7.33}$$

or

$$V_n = -\alpha G_0 M \frac{e^{-\mu a}}{a}, \tag{7.34}$$

to first order, while the force is

$$F_n = \frac{3\alpha G_0 M}{\mu^3 R^3}(\mu R \cosh \mu R - \sinh \mu R)(1 + \mu a)\frac{e^{-\mu a}}{a^2} \tag{7.35}$$

or

$$F_n = \alpha G_0 M(1 + \mu a)\frac{e^{-\mu a}}{a^2}, \tag{7.36}$$

to first order.

Summarizing the above results, we find that the non-Newtonian attraction outside a spherical mass has a potential similar to that of a point mass at the centre, but with a factor which depends upon the size of the mass; that factor is in practice very close to 1. When a solid sphere is used in an experiment to test the inverse square law, the size of the sphere does not have to be considered and only the distance between the attracting mass and attracted mass matters. On the other hand, the attraction inside a uniform shell, in contrast to the Newtonian attraction, does not vanish but is very nearly proportional to the distance from the centre. That can be used to make a direct test of non-Newtonian behaviour.

Non-Newtonian attraction of a cylinder

Thin ring

Consider first a thin ring of negligible thickness as a realistic model for attracting masses used in some experiments. As shown in Fig. 7.5, b is the radius of the ring, P is a point located on the axis of the ring at a distance y from the plane of the ring.

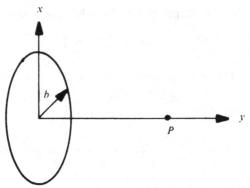

Fig. 7.5 Ring used as attracting mass in inverse-square law experiment.

The non-Newtonian potential at point P is

$$V_n = -\frac{\alpha G_0 M}{(b^2 + y^2)^{\frac{1}{2}}} \exp[-\mu(b^2 + y^2)^{\frac{1}{2}}], \qquad (7.37)$$

so that the non-Newtonian force is

$$F_n = \frac{\alpha G_0 My}{(b^2 + y^2)^{\frac{3}{2}}} \{\exp[-\mu(b^2 + y^2)^{\frac{1}{2}}] + \mu(b^2 + y^2)^{\frac{1}{2}} \exp[-\mu(b^2 + y^2)^{\frac{1}{2}}]\}. \qquad (7.38)$$

If it can be assumed that $1/\mu \gg (b^2 + y^2)^{\frac{1}{2}}$, then

$$F_n = \frac{\alpha G_0 My}{(b^2 + y^2)^{\frac{3}{2}}} [1 - \tfrac{1}{2}\mu^2(b^2 + y^2) + \ldots]. \qquad (7.39)$$

This formula will be applied below in the analysis of certain experiments.

Infinitely long hollow cylinder

As with a uniform spherical shell, the non-Newtonian gravitational force inside an infinitely long hollow cylinder may not vanish as the Newtonian force does.

Consider a semi-infinite hollow cylinder with internal and external radii R_1 and R_2 respectively (see Fig. 7.6). We again assume that

$$\mu R_i < 1 \quad (i = 1, 2) \qquad (7.40)$$

in the usual laboratory conditions.

The non-Newtonian part of the potential at a point P located in the basal plane at a distance a from the axis of the cylinder ($a < R_1$) is

$$V_n = -\alpha G_0 \rho \int_{R_1}^{R_2} dr \int_0^{2\pi} d\theta \int_0^\infty r \frac{\exp[-\mu(z^2 + r^2 + a^2 - 2ar\cos\theta)^{\frac{1}{2}}]}{(z^2 + a^2 + r^2 - 2ar\cos\theta)^{\frac{1}{2}}} dz; \qquad (7.41)$$

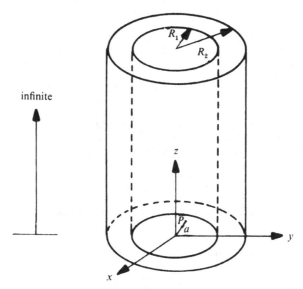

Fig. 7.6 Semi-infinite hollow cylinder.

where ρ is the density of the cylinder (assumed to be constant). On integration with respect to z, V_n becomes

$$V_n = -\alpha G_0 \rho \int_{R_1}^{R_2} dr \int_0^{2\pi} K_0(\eta)\, r\, d\theta, \qquad (7.42)$$

where

$$\eta = \mu(a^2 + r^2 - 2ar\cos\theta)^{\frac{1}{2}}, \qquad (7.43)$$

and $K_0(\eta)$ is the Basset (1886) function defined as

$$K_0(\eta) = -\ln\frac{\eta}{2}\sum_{k=0}^{\infty}\frac{\eta^{2k}}{2^{2k}(k!)^2} + \sum_{k=0}^{\infty}\frac{\eta^{2k}}{2^{2k}(k!)^2}\Psi(k+1). \qquad (7.44)$$

Here $\Psi(k+1)$ is the psi function,

$$\Psi(k+1) = -\gamma + \sum_{n=1}^{k}\frac{1}{n}, \qquad (7.45)$$

and γ is Euler's constant $(0.577\,215\,7\ldots)$.

Hence,

$$V_n = -\alpha G_0 \rho \int_{R_1}^{R_2} dr \int_0^{2\pi} r \left\{ -[\ln(\eta/2) + \gamma] \sum_{k=0}^{\infty} \frac{\eta^{2k}}{2^{2k}(k!)^2} \right.$$

$$\left. + \sum_{k=0}^{\infty} \frac{\eta^{2k}}{2^{2k}(k!)^2} \sum_{n=1}^{k} \frac{1}{n} \right\} d\theta. \quad (7.46)$$

The conditions $\mu R_1 < 1$ and $a < R$ ensure that

$$\eta < 1. \quad (7.47)$$

If we write

$$V_n = V_1 + V_2 + V_3, \quad (7.48)$$

then V_1, V_2 and V_3 have the following approximate values:

$$V_1 \approx \alpha G_0 \rho \int_{R_1}^{R_2} dr \int_0^{2\pi} r[\gamma + \ln(\eta/2)] d\theta, \quad (7.49)$$

$$V_2 \approx \tfrac{1}{4}\alpha G_0 \rho \int_{R_1}^{R_2} dr \int_0^{2\pi} r\eta^2 \ln\eta \, d\theta, \quad (7.50)$$

$$V_3 \approx -\tfrac{1}{4}\alpha G_0 \rho \int_{R_1}^{R_2} dr \int_0^{2\pi} r(1 - \gamma + \ln 2)\eta^2 \, d\theta. \quad (7.51)$$

Thus the radial non-Newtonian gravitational force on unit mass inside a semi-infinite tube is

$$F_n(\tfrac{1}{2}\infty) = -\tfrac{1}{2}\alpha\mu^2 G_0 \,\pi\rho(R_2^2 - R_1^2)(\ln 2 + \tfrac{1}{2} - \gamma)\,a$$

$$+ \tfrac{1}{2}\alpha\mu^2 G_0 \,\pi\rho(R_2^2 \ln \mu R_2 - R_1^2 \ln \mu R_1)\,a. \quad (7.52)$$

Using the superposition principle, the non-Newtonian gravitational force inside an infinitely long hollow cylinder will then be

$$F_n(\infty) = -\alpha\mu^2 G_0 \,Ha \left(\ln 2 + \tfrac{1}{2} - \gamma - \frac{R_2^2 \ln \mu R_2 - R_1^2 \ln \mu R_1}{R_2^2 - R_1^2} \right), \quad (7.53)$$

while H is the mass per unit length. Again the non-Newtonian, gravitational force inside an infinitely long cylinder will not be zero but will be proportional to the distance a from the axes.

Finite long hollow cylinder

In practice no cylinder can be infinitely long; we consider the force inside a cylinder of finite length, but long compared with the radii.

Fig. 7.7 shows a hollow cylinder with a finite length l. The radial non-

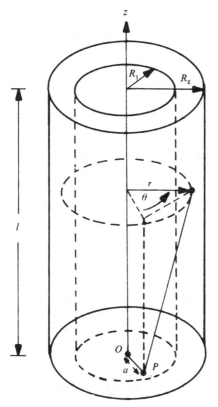

Fig. 7.7 Hollow cylinder of finite length.

Newtonian gravitational force on a unit mass located in the basal plane is given by

$$-\frac{\partial}{\partial a}\left\{-\alpha G_0 \rho \int_{R_1}^{R_2} dr \int_0^{2\pi} d\theta \int_0^l r \frac{\exp\left[-\mu(z^2+r^2+a^2-2ar\cos\theta)^{\frac{1}{2}}\right]}{(z^2+a^2+r^2-2ar\cos\theta)^{\frac{1}{2}}} dz\right\}.$$

The calculation of the above integral is not straightforward, but because the non-Newtonian gravitational force inside a semi-infinite hollow cylinder has already been obtained, $F_n(l)$ can be divided into two parts:

$$F_n(l) = F_n(\tfrac{1}{2}\infty) - \frac{\partial}{\partial a}\left\{-\alpha G_0 \rho \int_{R_1}^{R_2} dr \int_0^{2\pi} d\theta\right.$$

$$\left.\times \int_l^\infty r \frac{\exp\left[-\mu(z^2+r^2+a^2-2ar\cos\theta)^{\frac{1}{2}}\right]}{(z^2+a^2+r^2-2ar\cos\theta)^{\frac{1}{2}}} dz\right\}. \quad (7.54)$$

Let

$$R^2 = a^2 + r^2 - 2ar\cos\theta.$$

We then consider the integral

$$I = \int_l^\infty \frac{\exp[-\mu(z^2+R^2)^{\frac{1}{2}}]}{(z^2+R^2)^{\frac{1}{2}}} \, dz, \tag{7.55}$$

which because of the conditions $R/l < 1$, $\mu R < 1$ and $\mu l < 1$ becomes

$$I = -E_i(-\mu l) - \tfrac{1}{4}R^2\left[\mu^2 E_i(-\mu l) + \frac{\mu e^{-\mu l}}{l} + \frac{e^{-\mu l}}{l^2}\right] + \frac{3}{32}\frac{R^4}{l^4} + O\left(\frac{\mu^k R^4}{l^{4+k}}\right), \tag{7.56}$$

where $k \geqslant 2$, and $E_i(x)$ is the exponential integral

$$E_i(x) = \int_{-x}^\infty \frac{e^t}{t} \, dt. \tag{7.57}$$

A lengthy calculation then gives

$$\begin{aligned}
F_n(l) = &-\tfrac{1}{2}\alpha\mu^2 G_0\,\pi\rho(R_2^2 - R_1^2)(\ln 2 + \tfrac{1}{2} - \gamma)\,a \\
&+ \tfrac{1}{2}\alpha\mu^2 G_0\,\pi\rho(R_2^2\ln\mu R_2 - R_1^2\ln\mu R_1)\,a \\
&- \tfrac{1}{2}\alpha\mu^2 G_0\,\pi\rho(R_2^2 - R_1^2)\left[E_i(\mu l) + \frac{e^{-\mu l}}{\mu l} + \frac{e^{-\mu l}}{\mu^2 l^2}\right]a \\
&+ \tfrac{3}{8}\alpha G_0\,\rho\pi\frac{R_2^2 - R_1^2}{l^4}(a^2 + R_1^2 + R_2^2)\,a + \dots .
\end{aligned} \tag{7.58}$$

The non-Newtonian gravitational force at some position other than P in Fig. 7.7, can be found by the superposition principle as before. In particular, the force at a point O in the central horizontal plane of the cylinder is

$$F_n = 2F_n(\tfrac{1}{2}l), \tag{7.59}$$

and it can be see that once again it is proportional to the distance from the axis of the cylinder, the same result as that for an infinitely long cylinder, but with a different numerical factor depending on the dimensions of the hollow cylinder.

From the above study we may understand that the shape of an attracted test mass will also affect a non-Newtonian force upon it; however, since attracted test masses are usually very small, the effect is not taken into account and they are treated as point masses in most circumstances. If the attracted mass is a sphere, it can certainly be treated as a point mass.

7.5 Test of the inverse square law at the Cavendish Laboratory

We now give a detailed account of a test of the inverse square law carried out at the Cavendish Laboratory, Cambridge (Chen *et al.*, 1984). We use it to bring out the many points of design and particular implementation that are essential if one is to obtain results of high precision. We shall subsequently give brief accounts of other tests more or less contemporaneous with it.

Brief description of the experiment

The design of the experiment was such that the force exerted by a large cylinder on a test mass was compared with that exerted by a smaller cylinder closer to the test mass. The distances were chosen so that if the forces were Newtonian they would be equal, and any difference would correspond to a non-Newtonian component. The test mass was suspended from a torsion balance, and the net torque on the balance was reduced almost to zero by the attraction of a compensating cylinder placed on the opposite side of the test mass from that of the two other cylinders. In one group of experiments ('null' experiments) the net torque was made as small as possible, in a second group (non-null experiments) it was deliberately made different from zero.

Those two schemes were adopted, instead of just null experiments, because of a criticism that Long had made of the experiment of Spero *et al.* (1980). The Spero experiment was a null experiment, and Long had argued that it might for that reason have been insensitive to a vacuum polarization component of gravitation (Long, 1986). It was consequently arranged that there should be a definite non-zero field in the second set of experiments. It is doubtful if there is any substance in Long's argument; in fact, the two sets gave closely comparable results which, if that demonstrates anything, would seem to show that gravitational forces combine linearly.

The torsion balance had a light aluminium beam suspended by a tungsten fibre which had been subjected to the treatment described in Chapter 3. It was placed in a vessel at a pressure of 6×10^{-5} Pa, and was shielded from magnetic contamination and temperature fluctuations. The rotation of the torsion balance was detected by an optical lever in which light from an He–Ne laser was reflected from a mirror on the torsion arm to a position-sensitive photocell which was placed about 3 m from the torsion balance (Fig. 7.8).

The angular displacement of the balance was measured by an a.c. electronic system, which was chosen in preference to a d.c. system because of a higher signal-to-noise ratio and better zero stability. It is described later. The motion of the balance was damped by a voltage, which was proportional to the speed of the counterweight attached to the end of the balance arm, applied to the platinum plates P_1 and P_2 (see Fig. 7.8). This

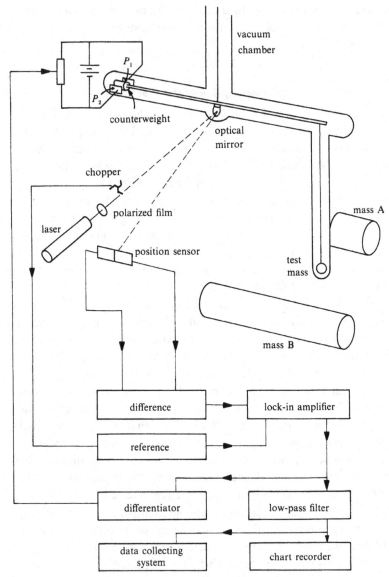

Fig. 7.8 Experiment of Chen *et al.* (1984) to test the inverse square law.

was necessary because the mechanical Q of the balance was extremely high $(10^3–10^5)$ and without electrical damping there would have been considerable difficulties in measuring the angular displacement of the beam. This scheme of electrical damping, as we mentioned in Chapter 2 does not introduce any additional force.

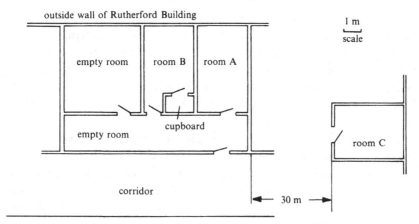

Fig. 7.9 Layout of the experiment of Chen *et al.* (1984) in the Cavendish Laboratory, Cambridge.

Location of the balance

The choice of the location of apparatus is very important in order to avoid disturbance by seismic motion and the gravitational attraction due to moving bodies other than the attracting masses.

The balance in this experiment was housed in the Rutherford Building of the Cavendish Laboratory in the University of Cambridge, which is about 3 km from the city centre and about 0.5 km from the nearest main road. Since any torsion balance is susceptible to changes in gravitational gradients in the horizontal plane, it should if possible be set up underground. Unfortunately, the Cavendish Laboratory has no basements and to minimize the effect of changing gravity gradients two precautions were taken: first, nearly all the observations were made in the early hours of the morning (2 a.m.–6 a.m.), because at that time of the day there was the least movement of people and apparatus in the building. Secondly, each deflection recorded was measured by an averaging method, which required readings of a given deflection to be taken at short intervals over a period of about half an hour. All local extraneous gravitational attractions that average out to zero on that time scale should then have no effect on the measurement.

The disposition of the rooms housing the balance and its associated instrumentation is indicated in Fig. 7.9. Room A contains the rotary and diffusion pumps which are required to pre-pump the vacuum chamber of the balance. To avoid the effect of the vibration they were switched off after the required vacuum had been reached and the reading could then be made. Room B contains a cupboard in which the torsion balance and associated attracting masses were housed. Both the cupboard and room B contain heat control units to ensure that the environmental temperature is kept reasonably constant. The temperature fluctuation over a 24 hour period amounted to about 0.5 deg. Room B also contains the optical bench, laser, position sensor and associated equipment. The proper functioning of the optical position sensor required this room to be blacked out.

To minimize the gravitational disturbance due to the observers, all instruments which were monitored, or from which readings had to be taken, were housed in room C situated some 30 m away from the rooms housing the torsion balance and associated equipment. This precaution was vital because the movement of people near the rooms housing the balance caused comparatively large deflections of the balance, and similar precautions are essential in all precise laboratory experiments on gravitation.

Structure of the torsion balance

The beam of the balance was about 0.6 m long and made of aluminium tubing with internal and external radii of about 3.30 mm and 6.35 mm respectively. It weighed (26 ± 0.6) g. The test mass was a phosphor bronze sphere weighing (41.027 ± 0.001) g with a diameter of 19.91 mm; its deviations from sphericity were less than 0.01 mm. It was hung from one end of the beam in a net of thin copper wire, 0.4 m below the beam.

It is always a problem in torsion pendulum experiments to attach the test mass to the beam for, if care is not taken, the ball could very easily be distorted to cause a systematic error. Hanging it in a thin net of copper wire proved to be a good solution and the copper wire connected the test mass electrically to earth.

A countermass weighing 42 g was attached to the opposite end of the beam; it was constructed from a hollow cylinder of brass and was fixed in position by a pure copper screw, brass being avoided because of its possible magnetization. The beam was adjusted to be horizontal by moving the counterweight to an appropriate position along the arm of the balance. All the components of the balance were cleaned ultrasonically before being sealed in the vacuum chamber.

Optical lever

The essential elements of an optical lever are a laser and a position-sensitive photocell. Both can cause uncertainties in the observations.

A laser as a light source gives a more stable intensity and a smaller angular drift of the beam than other light sources but none the less an ordinary unstabilized He–Ne laser showed changes of intensity of up to 10 per cent and angular shifts of up to 10^{-4} rad. The pointing instability is a particular problem for a precision experiment. Although the problems associated with intensity variations can be overcome by using a beam splitter to provide a reference beam, against which the intensity of the signal from the optical lever may be compared, angular drift cannot so readily be compensated.

It was therefore necessary to use a laser with the frequency stabilized for maximum intensity in the fundamental mode and to suppress the high order modes which cause lateral beam jitter. The intensity of the laser used in the experiment was stable to 0.1 per cent and the angular drift was less than 5×10^{-8} rad when operated in room B with a daily temperature fluctuation of not more than 0.5 deg.

For the torsion balance to achieve a sensitivity of 5×10^{-14} m s^{-2}, the angular sensor must have a resolution of about 2×10^{-9} rad, as can be seen from eqn (4.62).

The position-sensitivity of the photocell was about 0.1 nm. Thus for a signal-to-noise ratio of 100 the lever arm of the light beam should be about 3 m, a reasonable value which was in fact used. The total length of the active area of the photocell was 3 cm, corresponding to a beam displacement of 5×10^{-3} rad and to a motion of the counterweight at the end of the arm of 1.5 mm. That established the spacing of the platinum electrodes, P_1 and P_2 (Fig. 7.8) by which feedback was applied.

The position-sensitive photocell suffered from two drawbacks. The first was thermal drift of the null point, which was typically of the order of 1 μm per 10^{-2} deg, or in terms of the angular displacement of the torsion balance, 3×10^{-7} rad deg^{-1}. It was not too serious in the present experiment because it was always possible to ensure that the duration of a single measurement was much shorter than the drift time of the sensor, typically of the order of 24 hours. A more significant problem was the relation between the signal from the sensor and the true angular displacement. The positional response of the photocell was investigated with great care and it was found that the maximum deviation from a linear response was about 1.3 per cent over the whole operating range. That is of no consequence for a null experiment because the aim is to determine whether or not there is any beam deflection at all. If, however, the forces are not balanced and there is a deflection, the non-linear behaviour of the device can have an effect, and the angle in eqn (4.62) must be calibrated. Unfortunately,

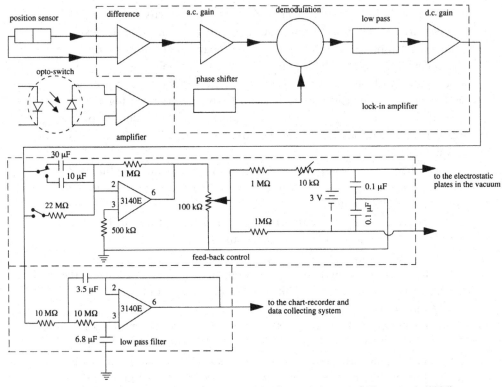

Fig. 7.10 Electric control circuits used in the experiment of Chen *et al.* (1984).

observations showed that the non-linearity is somewhat irregular and not easy to calibrate over the whole range. Therefore, in order to achieve an uncertainty in the test of the inverse square law of less than 1 part in 10^4, the useful range of the sensor had to be limited to less than 1 per cent of the whole operational range. Consequently servocontrol by proportional feedback was applied to keep the dynamical range of deflection within suitable limits.

Amplifier and servo-system

A block diagram of the electronics is given in Fig. 7.10. The difference signal obtained from the position sensor is rather weak, particularly because the laser light is reduced with a polarizer in order to minimize the effect of the fluctuation of light pressure on the mirror of the beam. To avoid the d.c. drift associated with a d.c. amplifier, the laser light was chopped to provide an amplitude-modulated signal at about 200 Hz. The signal was then amplified by a precision lock-in amplifier with a gain of about 10^6.

The necessity of incorporating an appreciable amount of damping in the torsion balance has been described in the early part of this book; it was provided by the servo-system. Two types of feedback were used: one was a simple differential feedback to provide electrical damping proportional to the speed of the counterweight of the balance and the other was a combination of negative proportional feedback and differential feedback which was used in the non-null experiment to limit the dynamical range of swing of the torsion balance. The feedback signals were applied to the two electrodes P_1 and P_2 enclosed inside the vacuum chamber. In order to linearize the feedback signal, a bias voltage was applied between the plates, as explained in section 4.9.

Electrical and magnetic shielding

The importance of shielding the torsion balance from magnetic and electrical contamination has been discussed in Chapter 3.

The vacuum chamber was made of glass, and was wrapped with aluminium foil to provide an electrostatic shield that was earthed by a thick copper wire to a water pipe. The moving parts of the balance (i.e. the beam, the test mass and the counterweight) were earthed through the tungsten wire and the rotatable head of the balance.

The problem of magnetic shielding is more difficult but it was possible to reduce magnetic effects to a minimum. Such precautions are important, for it will be recalled from Chapter 3 that a magnetic dipole moment as small as 10^{-3} G cm (10^{-9} T m) acting in the Earth's field would produce an acceleration of the balance of 10^{-11} m s^{-2}. First, all materials used in the construction of the balance were either diamagnetic or paramagnetic. Secondly, sufficient magnetic shielding was employed. The most important part of the balance to shield is the test mass, and the lower end of the vacuum chamber, in which the test mass was housed, was placed within a cylindrical mumetal box (Fig. 7.11). The effectiveness of this shielding was not easy to gauge accurately, but some rather rough measurements involving the movement of a small magnet in the vicinity of the test mass showed that one can expect stray field strengths to be attenuated by a factor of about 15 or more and in practice it seems that was sufficient. The effectiveness of the shielding was probably limited by the low ratio of t/d (eqn (3.30)).

Thermal shielding

The serious effects of drifts of temperature were pointed out in earlier chapters, a dramatic demonstration of which was shown by switching on

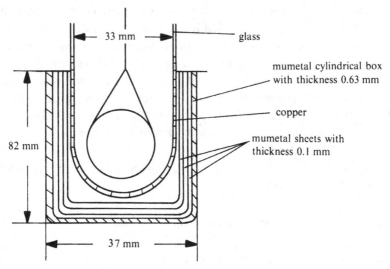

Fig. 7.11 Magnetic shielding of test mass.

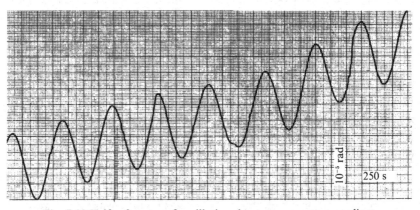

Fig. 7.12 Drift of centre of oscillation due to temperature gradient.

a small electrical fire next to one wall of the cupboard in which the balance was housed. A record of the behaviour of the balance is shown in Fig. 7.12, where a very strong drift is observed.

To reduce the effects of temperature on the behaviour of the balance, the following precautions were taken:

(1) To reduce the effect of residual stress in the tungsten fibre caused by temperature change, the fibre was treated by the process described in section 3.5 (see Table 3.2).

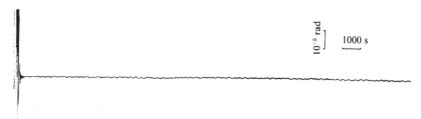

Fig. 7.13 Record of torsion balance rotation.

(2) A high vacuum was maintained in the balance chamber to minimize the heat flux passing through the moving parts of the balance.

(3) The lower part of the vacuum vessel surrounding the test mass was constructed of copper (sealed to the glass tube). The high thermal conductivity of the copper ensured that temperature gradients in this vital region of the balance were kept low.

(4) The entire vacuum chamber was surrounded by a box of aluminium, some 0.8 m in length and 0.2 m in width and the height and the space between it and the vacuum chamber was filled with granules of polystyrene foam. The box acted as a thermal filter attenuating the temperature fluctuations of the room.

(5) The room temperature was controlled by heaters, which kept it constant to within 0.5 deg per day.

The effect of these precautions is illustrated in Fig. 7.13 which shows a recording of the movement of the torsion balance with a slight electrical damping over 6 hours. Long-term observations have shown that the stability of the balance was typically 2×10^{-6} rad per day, or 10^{-7} rad per hour.

Attracting masses

The attracting masses were right circular cylinders fabricated from non-magnetic stainless steel. The dimensions and masses of the three cylinders, identified by the letters A, B and C, are given in Table 7.2. Systematic errors may arise from two possible sources: irregularities in the cross-sections of the cylinders and inhomogeneities of the density.

The cross-sections of the cylinders were examined for circularity with a Talyrond instrument and the maximum departure of any of the cylinders from circularity was found to be 0.5 μm corresponding to an error of less than 5×10^{-6} in the final result of the test of the inverse square law, as calculated from the exact solution for the gravitational attraction of a cylinder.

Table 7.2. *The attracting masses in the experiment of Chen et al.* (*1984*)

Cylinder	Length (mm)	Diameter (mm)	Mass (kg)
A	100.007 ± 0.001	100.000 ± 0.001	$6.251\,331 \pm 1 \times 10^{-5}$
B	399.997 ± 0.005	99.996 ± 0.001	$25.000\,07 \pm 4 \times 10^{-5}$
C	100.0026 ± 0.001	99.998 ± 0.001	$6.250\,565 \pm 1 \times 10^{-5}$

Table 7.3. *Densities of the cylinders*

Cylinder	Density (kg m^{-3})
A	7958.80
B	7958.31
C	7958.47

The inhomogeneity of density is much more troublesome. The density distribution could not be examined directly, but, following the method used by Heyl (1930), a check was conducted of the homogeneity of the cylinders by determining their average densities from measurements of their dimensions and masses. The results obtained are given in Table 7.3.

This is a gross test which suggests that the density inhomogeneity is of the order of 5 parts in 10^5. Nevertheless the precaution was taken of rotating each cylinder about its axis into a different position for each of 40 measurements to average out any effect of the density inhomogeneities in the radial direction.

The construction of the bench that was designed to hold the attracting masses is shown in Fig. 7.14. Each cylinder was held in position by means of two rods mounted on a 2 cm thick aluminium plate. Three micrometers, indicated by 1, 2 and 3 in Fig. 7.14, were used to traverse the table vertically, while micrometer 4 was used to traverse the table horizontally. An accurate control of the position of the whole table in the x-direction of Fig. 6.5 was essential for the success of the procedure used to find the mid-position of the two cylinders A and C, which were supposed to be identical. The alignment of the attracting masses with the test mass in the y-direction (also see Fig. 6.5) was far less crucial in determining the accuracy of the experiment, as was discussed in Chapter 6. It was achieved with the aid of a fine mark indicating the middle position of the cylinder along the long

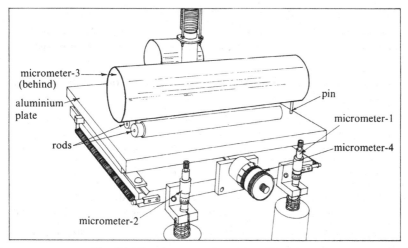

micrometer-3 (behind)

aluminium plate

rods

micrometer-2

pin

micrometer-1

micrometer-4

Fig. 7.14 Bench supporting attracting masses.

axis and also of the copper wire hung from one arm of the balance, from which the test mass was hung. Once the correct position of the attracting masses in the y-direction was found, pins were inserted in the table to indicate their end position (see Fig. 7.14), so allowing the removal of the cylinders and their subsequent replacement without resorting to the initial alignment procedure.

Measurement procedure and data collection

The measurement procedures were as follows.

(a) The distance between cylinders A and C and between cylinders A and B must be measured as accurately as possible. This was achieved by using a set of high precision gauge blocks with an overall tolerance of 0.1 μm for the set.

(b) With the attracting masses far removed from the test mass of the torsion balance, the optical lever was adjusted so that the laser beam reflected from the mirror of the balance was incident roughly in the centre of the position sensor. The output of the lock-in amplifier V_1 was then measured.

(c) With cylinders A and C in position on the table the output voltage V_2 was measured.

(d) Micrometer 4 in Fig. 7.14 was then adjusted by a small amount and steps (b) and (c) were repeated until eventually the voltages V_1 and V_2 were equal.

The steps (b)–(d) were necessary to set the test mass in the right position relative to the attracting masses and ensured that the distance between the test mass and cylinder A would be half the distance between A and C.

(e) With masses A and B only on the table, the output voltage V_{AB} was measured.

(f) With masses A and C only on the table, the output voltage V_{AC} was measured.

(g) With mass A only in position on the table, the output voltage V_A was measured.

The difference $V_{AB} - V_{AC}$ would represent the torque difference due to the attracting masses B and C at different distances, which is very small; while $V_A - V_{AC}$ would represent the absolute torque produced by mass C. Then the calculated quantity

$$\left(\frac{\Delta\tau}{\tau}\right)_{\text{exp}} = \frac{V_{AB} - V_{AC}}{V_A - V_{AC}}, \tag{7.60}$$

is a relative deviation to be compared with the corresponding theoretical result.

Steps (e), (f) and (g) are those involved in the actual test of the inverse square law. Step (g) is equivalent to making a calibration for the voltage measurement and steps (e) and (f) were repeated many times to generate sufficient data to allow an accurate estimate of the mean of the readings. Each time the procedures were repeated, the cylinders were rotated into a new position to reduce any influence due to the density inhomogeneities.

The cylinders, when removed from the table as required in the above procedures, were not taken out of Room B, because it was important that they should remain at the same temperature as the balance. Initially the cylinders were indeed removed from the room in order to reduce their gravitational influence; but when they were returned to the table, the small temperature differences developed as a result of their removal caused some drift of the zero point of the balance. To avoid this, the cylinders, when removed from the table, were placed in preset positions in the same room, in line with the arm of the balance so that they did not produce a torque upon it.

Further problems may arise from people moving in and out of the experimental room to exchange the attracting cylinders. The room temperature may be disturbed, the balance may be attracted by the mass of the person who enters the room, the optical sensor may be affected because

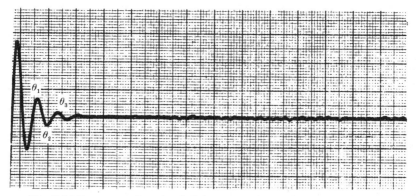

Fig. 7.15 Response of torsion balance to exchange of attracting masses.

lamps in the room had to be switched on. It would seem that an attracting mass should be moved automatically but that would create other problems, particularly with heavy masses: for example, local vibration and tilting caused by moving the heavy mass, disturbance of the stray magnetic field by a motor, position accuracy of the moving mechanism, and so on. Manual changing of the masses caused the torsion balance to start to swing but it was found that if it could be done skilfully and promptly, along with a proper damping of the balance, readings could be taken with confidence after a reasonable time. With experience an assistant could enter the room, exchange the masses in a fairly regular way and then leave the room within about 8 to 10 seconds and with minimum disturbance. The next reading could be made after about half an hour.

It has already been seen that it is important to collect the data for a given experimental run within a limited time interval to minimize the effect of any drift of the equilibrium position of the balance. On the other hand, the balance must have time to recover from the disturbance caused by changing the attracting masses.

Two methods have been used.

(1) *Graphical method (balance slightly damped).* As the solution of the differential equation to describe the oscillatory behaviour is

$$\theta = A_0 + A\,e^{-\beta t}\cos\omega t, \tag{7.61}$$

the value of the equilibrium position A_0 can be found from three successive extreme values of the deflection, θ_1, θ_2 and θ_3, as noted in Fig. 7.15:

$$A_0 = \left(\frac{\theta_1\theta_2 - \theta_2^2}{\theta_1 + \theta_3 - 2\theta_2}\right)^{\frac{1}{2}}. \tag{7.62}$$

The advantage of this method is that the balance does not need to be totally stopped before the reading can be taken.

(2) *Averaging method (balance critically damped).* The critically damped balance was left for half an hour to allow it to come to rest at the new equilibrium position. The displacement of the balance was then read twice a second for the next 30 minutes yielding a total of 3600 readings which were automatically averaged by a digital voltmeter to determine an equilibrium position of the balance.

Tests over a long time showed that for the same numbers of measurements the graphical method yielded a slightly larger standard deviation of an individual measurement. However, it is believed that both the methods should be equally accurate. The difference is thought to be due to the effects of noise, Brownian motion, seismic motion and external impulses upon the motion of the balance, which were to some extent reduced by the averages taken in the second method; that was accordingly adopted for the results.

Experimental results from the measurements

For the non-null experiment

This is the experiment in which the torque on the test mass is not zero. The separations of the attracting masses for the non-null experiment are shown in Fig. 7.16. Altogether 43 measurements were made using the procedure described above. The result was

$$\left(\frac{\Delta\tau}{\tau}\right)_{\text{non-null}} = (6.691 \pm 0.011) \times 10^{-2}. \tag{7.63}$$

For the null experiment

This is the experiment in which the torque on the test mass is roughly zero. The separations of the attracting masses for the null experiment are shown in Fig. 7.17. Again, 43 measurements were made using the procedure described above. The result was

$$\left(\frac{\Delta\tau}{\tau}\right)_{\text{null}} = (1.438 \pm 0.018) \times 10^{-2}. \tag{7.64}$$

The mean of the measurements is not exactly zero because the relative position of the attracting masses was preset. The reason that we did not design this experiment as an exact null experiment will be explained below.

unit: mm

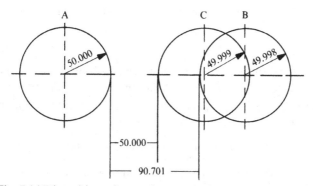

Fig. 7.16 Disposition of attracting masses in non-null experiment.

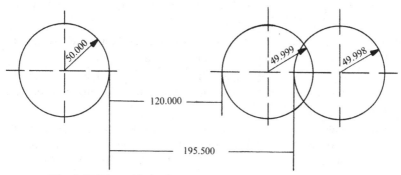

Fig. 7.17 Disposition of attracting masses in null experiment.

The standard deviation of the non-null experiment was less than that of null experiment because in the null experiment the distances of the attracting masses were greater than in the non-null experiment so that the gravitational attractions were weaker and random uncertainties relatively greater.

Theoretical results from the inverse square law and the final results

For the non-null experiment

To determine whether or not the above experimental results can be accounted for by purely Newtonian forces, a complete calculation of those forces must be carried out that will include all gravitational attractions, not just the attraction between the test mass and the attracting mass but also the attractions between all the moving parts of the balance and the attracting masses. Imperfections of the experimental arrangements must also be examined in detail. The contributions to the observed torque are as follows.

(1) The term $(\tau_C - \tau_B)/\tau_C$, denoted as $\delta_{nn}^{(1)}$, where τ_B and τ_C are the Newtonian torques between the test mass and the attracting masses B and C respectively. This term is the main contribution to the final result; the forces F_C and F_B, according to Chapter 6 and the dimensions in Fig. 7.16, are respectively

$$F_C = 4G_0 \rho C_y(50.0013, 49.999, 75.000) \qquad (7.65)$$

and

$$F_B = 4G_0 \rho C_y(199.9985, 49.998, 115.699), \qquad (7.66)$$

where $C_y(L, R, a)$ is the function defined in eqn (6.18).

(2) The difference in the Newtonian gravitational attraction between the counterweight and the masses C and B respectively needs to be accounted for. This difference is denoted as $\delta_{nn}^{(2)}$.

(3) Similarly, the difference in the Newtonian gravitational attractions between the arm of the balance and the masses C and B respectively needs to be calculated. This difference is denoted as $\delta_{nn}^{(3)}$ and the dimensions of the apparatus required for the calculation of $\delta_{nn}^{(3)}$ are given in Fig. 7.18.

(4) Because the apparatus was not perfectly set up, a correction must be made in order to take into account the fact that the axes of the attracting cylinders are not exactly in the same plane. This difference was measured using a cathetometer, which showed that mass B was 0.42 ± 0.01 mm lower than the level of masses A and B. This correction is denoted as $\delta_{nn}^{(4)}$.

(5) Although the masses A and C were intended to be identical, they were not so. The correction due to the small difference of mass between A and C amounts to about 1 part in 10^4 and the zero position found by steps (b) and (c) of the measurement procedure did not coincide exactly with the true centre. This correction is denoted as $\delta_{nn}^{(5)}$.

(6) Because this was a non-null experiment, the equilibrium position of the test mass altered slightly according to whether mass B or mass C was present on the table during steps (e) and (f) of the measurement procedure. The resulting shift in the zero position of the test mass amounted to 22 µm and led to a small correction which is denoted as $\delta_{nn}^{(6)}$. The correction made here is an example of one that it is easy to overlook.

Table 7.4. *Newtonian contributions to the net force – non-null experiment*

Significance	Items	Relative figures
Exp nominal difference	$\delta_{nn}^{(1)}$	6.2205×10^{-2}
Attraction of counterweight	$\delta_{nn}^{(2)}$	0.7011×10^{-2}
Attraction of arm	$\delta_{nn}^{(3)}$	-0.2687×10^{-2}
Cylinders not in same plane	$\delta_{nn}^{(4)}$	0.0015×10^{-2}
Cylinders not identical	$\delta_{nn}^{(5)}$	0.0032×10^{-2}
Test mass not central	$\delta_{nn}^{(6)}$	0.0221×10^{-2}
	Total	6.6797×10^{-2}

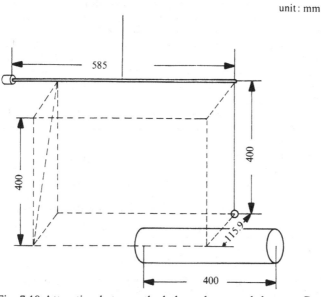

unit: mm

Fig. 7.18 Attraction between the balance beam and the mass B.

The calculation of the six terms is tedious, particularly that for the correction of the arm of the torsion balance, but it is necessary to consider them all thoroughly. The relative values of $\delta_{nn}^{(i)}$ are given in Table 7.4.

The results of the calculation listed above were based on the nominal dimensions and distances of the experimental arrangements. As a matter of fact, every figure used in the calculation has its own measurement error; therefore the accuracy with which the theoretical Newtonian value of $\Delta\tau/\tau$ can be calculated is limited by uncertainties about the various parameters of the experiment. These parameters, together with their errors, are listed

Table 7.5. *Components of error – non-null experiment in the calculation of the theoretical results*

Parameters	Uncertainty	Error contribution
Distance between cylinders	A and C (50.000 ± 0.001) mm	$\pm 0.0029 \times 10^{-2}$
	A and B (90.701 ± 0.001) mm	$\pm 0.0017 \times 10^{-2}$
Diameters of cylinders	B: (99.996 ± 0.001) mm	$\pm 0.002 \times 10^{-2}$
	C: (99.998 ± 0.001) mm	$\pm 0.002 \times 10^{-2}$
Lengths of cylinders	B: (399.997 ± 0.005) mm	$\pm 0.0013 \times 10^{-2}$
	C: (100.0026 ± 0.001) mm	$\pm 0.001 \times 10^{-2}$
Non-circularity of cylinders	0.0005 mm	$< 0.0005 \times 10^{-2}$
Density inhomogeneity of cylinders	5×10^{-5}, but with rotation of the cylinders may be reduced by factor of 2–3	$\pm 0.002 \times 10^{-2}$
Central position of test mass	$\Delta x = \pm 0.004$ mm	$\pm 0.004 \times 10^{-2}$
	$\Delta y = \pm 0.05$ mm	$< 0.0001 \times 10^{-2}$
	$\Delta z = \pm 0.01$ mm	$< 0.0001 \times 10^{-2}$
Non-linearity of detection system		$\pm 0.004 \times 10^{-2}$
	Total	$\pm 0.0076 \times 10^{-2}$

in Table 7.5. Therefore, the value of $\Delta\tau/\tau$ as calculated from Newtonian theory is

$$(\Delta\tau/\tau)_{\text{theo}} = (6.680 \pm 0.0076) \times 10^{-2}, \tag{7.67}$$

whereas the observed value was $(6.691 \pm 0.011) \times 10^{-2}$, and thus

$$(\Delta\tau/\tau)_{\text{nn}} - (\Delta\tau/\tau)_{\text{theo}} = (1.1 \pm 1.3) \times 10^{-4}. \tag{7.68}$$

For the null experiment

The various corrections for the null experiment are denoted by $\delta^{(i)}_{\text{nu}}$ and calculated values are listed in Table 7.6. It can be seen that the null experiment was well designed with respect to the main term $\delta^{(1)}_{\text{nu}}$, but the attraction of the counterweight was considerable and turned out to be the main contribution to an overall effect that was not entirely null.

As before, the accuracy with which the Newtonian force can be calculated was estimated, with the result that

$$(\Delta\tau/\tau)_{\text{theo}} = (1.430 \pm 0.009) \times 10^{-2} \tag{7.69}$$

as compared with the observed value $(1.438 \pm 0.018) \times 10^{-2}$ giving, for the null experiments,

$$(\Delta\tau/\tau)_{\text{null}} - (\Delta\tau/\tau)_{\text{theo}} = (0.8 \pm 2.0) \times 10^{-4}. \tag{7.70}$$

Table 7.6. *Newtonian contributions to the net force – null experiment*

Significance	Items	Relative figures
Exp nominal difference	$\delta_{nu}^{(1)}$	0.0064×10^{-2}
Attraction of counterweight	$\delta_{nu}^{(2)}$	2.1256×10^{-2}
Attraction of arm	$\delta_{nu}^{(3)}$	-0.7056×10^{-2}
Cylinders not in same plane	$\delta_{nu}^{(4)}$	0.0007×10^{-2}
Cylinders not identical	$\delta_{nu}^{(5)}$	0.002×10^{-2}
Test mass not central	$\delta_{nu}^{(6)}$	0.001×10^{-2}
	Total	1.4301×10^{-2}

Interpretation of the experimental results

Interpretation of the non-null experiment

The non-Newtonian forces F_{CnN} and F_{BnN} produced by cylinders C and B were computed by numerical integration over the whole volume with the results, for cylinder C,

$$F_{CnN} = 423.63\alpha\mu^2 G_0 \rho \times 10^{-6} \text{ m}^3, \qquad (7.71)$$

and for cylinder B,

$$F_{BnN} = 1483.75\alpha\mu^2 G_0 \rho \times 10^{-6} \text{ m}^3. \qquad (7.72)$$

Thus, the difference between them that would be expected in the experiment is

$$\Delta F_{nN} = 1060.12\alpha\mu^2 G_0 \rho \times 10^{-6} \text{ m}^3. \qquad (7.73)$$

That non-Newtonian component should be divided by the gravitational force F_C, produced by cylinder C, namely

$$F_C = 12.5842 G_0 \rho \times 10^{-2} \text{ m}, \qquad (7.74)$$

to give

$$\frac{\Delta F_{nN}}{F_C} = 84.24\alpha\mu^2 \times 10^{-4} \text{ m}^2, \qquad (7.75)$$

which is to be compared with the limit of eqn (7.68), namely

$$\left(\frac{\Delta\tau}{\tau}\right)_{nn} - \left(\frac{\Delta\tau}{\tau}\right)_{theo} = \frac{\Delta F_{nN}}{F_C} = (1.1 \pm 1.3) \times 10^{-4}. \qquad (7.76)$$

Then eqn (7.75) and (7.76) give the value of the product $\alpha\mu^2$ as

$$\alpha\mu^2 < 0.028 \text{ m}^{-2} \qquad (7.77)$$

or, if $\alpha = \frac{1}{3}$, the lower limit of the Compton wavelength is

$$\mu^{-1} > 3.4 \text{ m}. \qquad (7.78)$$

Interpretation of the null experiment

Following the same procedure the relative magnitude of the non-Newtonian force in the null experiment is

$$\frac{\Delta F_{nN}}{F_C} = 186.69\alpha\mu^2 \times 10^{-4} \text{ m}^2 \tag{7.79}$$

a value much larger than that of eqn (7.75); the reason is that the masses were set further away from the test mass in the null experiment, so that the Newtonian attraction of mass C was smaller.

Then, as before, the value for the product $\alpha\mu^2$ can be calculated as

$$\alpha\mu^2 < 0.014 \text{ m}^{-2} \tag{7.80}$$

or, if $\alpha = \frac{1}{3}$, the lower limit of the Compton wavelength is

$$\mu^{-1} > 4.9 \text{ m}. \tag{7.81}$$

Here we can see that the null experiment gives a slightly better limit than the non-null experiment, although it is less precise because of the weaker signal produced by the longer distance from the test mass. That shows the advantage of using the $\alpha\mu^2$ analytical method for it depends upon the distance as well as upon the precision of the experiment. Some experiments with poorer precision at a longer distance may achieve a better result.

Some remarks on the interpretation of the experimental results using the $\alpha - \mu^{-1}$ plot

The formula (eqn (7.16)) by which the boundaries of region in the (α, μ^{-1})-plane are found, applies only for point masses, and when account is taken of the extension of the attracting mass, the formula could become very complicated. We therefore, introduce the concept of 'equivalent separation', namely the separation between the test mass and an imagined attracting point body with the same mass, which is assumed to be placed in such a position that it produces the same Newtonian attraction as that of the real physical attracting mass. It is easy to calculate that in the non-null experiment the two equivalent separations were

$$r_1 = 75 \text{ mm} \quad \text{(for mass A)},$$
$$r_2 = 170 \text{ mm} \quad \text{(for mass C)},$$

while in the null experiment they were

$$r_1 = 110 \text{ mm} \quad \text{(for mass A)},$$
$$r_2 = 250 \text{ mm} \quad \text{(for mass C)}.$$

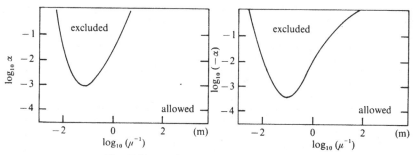

Fig. 7.19 α–μ^{-1} plot for non-null experiment.

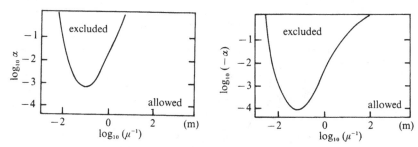

Fig. 7.20 α–μ^{-1} plot for null experiment.

The boundaries in the (α, μ^{-1})-plane corresponding to the measured value

$$G(r_1)/G(r_2) = 1 + (1.1 \pm 1.3) \times 10^{-4}$$

of the non-null experiment are shown in Fig. 7.19.

Similarly, the boundaries corresponding to the measured value

$$G(r_1)/G(r_2) = 1 + (0.6 \pm 2.0) \times 10^{-4}$$

in the null experiment are shown in Fig. 7.20.

7.6 Long's experiment

In 1976, Long (1976) claimed that the inverse square law of gravitation breaks down at a distance of the order of 0.1 m. He had earlier made a detailed analysis (Long, 1974) of the measured values of G, as reported in the literature from 1884 to 1969, in which he paid attention to the intermass separations used by various authors who made measurements of G during the first half of the twentieth century; the analyses seemed to indicate that the value of G depended on the intermass separations.

In his own experiment, Long (1976) compared the value of the gravitational constant at a mass separation of 299 mm with the value at a

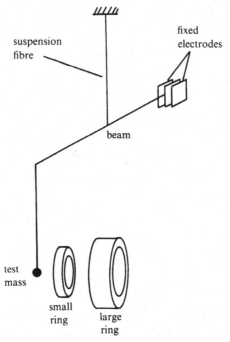

Fig. 7.21 Apparatus of Long (1976).

mass separation of 44.8 mm using a torsion balance. The attracting masses were chosen so that the forces exerted were the same according to the inverse square law. Long dispensed with the traditional spherical shape for the attracting masses in favour of masses in the shape of rings or hollow cylinders; his reason for using rings was that, as discussed in Chapter 6, the direct measurement of the separation of the attracting mass from the attracted mass is not required, because of the mathematical behaviour of the gravitational attraction around the stationary point of a ring (see Fig. 6.10). The experiment was performed, as shown in Fig. 7.21, in such a way that the rings were placed in turn to exert their maximum attraction upon the test mass on the torsion balance. The forces required to annul the deflection of the torsion balance were the quantities to be measured in order to study the possible deviation from the inverse square law.

The larger ring was made of brass and weighed 57.58 kg. It was 76.3 mm thick and the inner and the outer radii were 215.9 mm and 271.1 mm respectively. The smaller ring was of tantalum and weighed 1.22 kg. It was 17.8 mm thick and the inner and outer radii were 27.5 mm and 45.5 mm respectively. Long mentioned in his paper that the stationary point was

174.15 mm from the face near the test mass, while that of the small ring was 26.088 mm. However, it is believed that these values are misprints. A calculation using eqn (6.40) shows that the distances of the stationary points for the large and small rings should be 135.98 mm and 17.217 mm respectively. The test mass was a sphere of tantalum which weighed 50 g and was suspended 870 mm below the arm of the torsion balance.

The arm of the torsion balance was suspended by a tungsten fibre 50 μm in diameter and 486 mm long. The balance was in an enclosure evacuated to 10^{-5} Pa, with the wall temperature constant to within 0.01 deg. The deflection of the balance was observed with an optical lever – a mirror reflected a beam of light onto an array of photodetectors arranged as a position detector.

Great care was taken to reduce the vibration of the torsion arm and the top of the tungsten fibre was suspended from a damped vibration isolator.

The results were expressed in terms of the relative difference of the torques produced by the rings. If T_1 is the torque exerted by the large ring and T_s is that exerted by the small ring, the relative difference is equal to

$$\frac{T_1 - T_s}{T_s}.$$

When the departure from the nominal dimensions and masses of the rings is taken into account and the forces of the rings on the arm of the balance are calculated, the relative difference of the torque on the assumption of an inverse square law would be 0.03807; however, the mean value found experimentally was 0.0417 ± 0.0007. Therefore the difference is 0.0037 ± 0.0007, which was thought by Long (1976) to be the deviation from the Newtonian law.

To interpret this experimental result, we go back to the previous study of the non-Newtonian force of the laboratory masses. For simplicity, we treat the ring as an ideal ring without thickness, so that eqn (7.39) can be used and the relative change of G can be expressed as

$$\frac{\Delta G}{G} = -\tfrac{1}{2}\alpha\mu^2(b_2^2 - b_1^2 + d_1^2 - d_2^2), \tag{7.82}$$

where b_1, b_2 are the radii of the two rings, and d_1, d_2 are the distances of the test mass from the rings. Inserting Long's data into the above equation, we have

$$0.07 < \alpha\mu^2 < 0.1 \text{ m}^{-2} \tag{7.83}$$

or, if $\alpha = \tfrac{1}{3}$ is chosen, the wavelength of the assumed exchange particle is

$$1.8 \text{ m} < \mu^{-1} < 2.2 \text{ m}. \tag{7.84}$$

The corresponding α–μ^{-1} plot is shown in Fig. 7.25.

It is still not known why Long's result was different from other measurements. However, the following comment may be useful.

In all experiments to measure G or to test the inverse square law (unlike those to test the principle of equivalence) the main error lies in the metrology, in particular the measurement of the separation between the attracting mass and the attracted mass. As has been seen, there are various ways of designing experiments to avoid direct measurement and one way is to use a hollow cylinder with its stationary point on the axis, as Long did. The mathematical behaviour of the gravitational field of a hollow cylinder around the stationary point has been discussed in Chapter 6, where the gravitational field was seen to be a hyperparaboloidal surface in a four-dimensional space. That means that although it has a maximum value along the z-axis, it has a minimum along any radius. It seems to us, from Long's publications, that he may not have appreciated that the stationary point was a saddle point, not an absolute maximum. If the test mass were located only at a maximum along the z-axis, without making sure that it was also at the minimum point on the other two axes, it would be possible for a systematic error have occurred.

7.7 First experiment of Newman and co-workers

Spero *et al.* (1980) carried out an experiment analogous to the Faraday cage in electromagnetism. If the inverse square law holds, then the force inside a closed cavity, and in particular that inside an infinitely long hollow cylinder, should be zero. On the other hand, as demonstrated by eqn (7.53) the non-Newtonian force at any point inside an infinitely long hollow cylinder is not zero, it is proportional to the product of α and μ^2, the distance a from the central line of the cylinder and the mass density per unit length. That affords the possibility of measuring the non-Newtonian force in a direct way.

It is of course, not possible to make an infinitely long cylinder. However, as has already been seen (Chapter 6), the radial component of the Newtonian force inside a finite long hollow cylinder at a distance a from the axis is proportional to a/L^3, so that in a sufficiently long cylinder the Newtonian force will be insignificant. Further, in the experiment of Spero *et al.* that residual Newtonian force was cancelled by the Newtonian gravitational force of a thin ring. We now show that although the Newtonian force is cancelled, the non-Newtonian force will remain.

As we know, the sum of the Newtonian and non-Newtonian forces at a point on the axis of a ring and at a distance d from the plane of the ring is

$$F_r = (1+\alpha)\frac{G_0 M_r d}{(b^2+d^2)^{\frac{3}{2}}} - \tfrac{1}{2}\alpha\mu^2 \frac{G_0 M_r d}{(b^2+d^2)^{\frac{1}{2}}} + \dots, \tag{7.85}$$

provided that the thickness of the ring may be ignored; b is the radius of the ring, M_r is the mass of the ring.

The total force at a on the mid-plane inside a hollow cylinder, is

$$F_h = -(1+\alpha)\frac{4G_0 M_h a}{L^3}\left[1 - \frac{3a^2}{L^2} - \frac{3(R_2^2 + R_1^2)}{L^2} - \dots\right] + \alpha\mu^2 \frac{G_0 M_h a}{L}$$
$$\times\left[\ln 2 - \gamma + E_i(-\tfrac{1}{2}\mu L) - \frac{R_2^2 \ln \mu R_2 - R_1^2 \ln \mu R_1}{R_2^2 - R_1^2} + \dots\right], \tag{7.86}$$

where a is the radial distance of the test mass, and M_h and L are the mass and length of the cylinder. Clearly, from the form of the two equations, if the Newtonian components cancel each other, then because of the common factor $(1+\alpha)$ the non-Newtonian component proportional to just the parameter α by itself will also be cancelled. Thus, the residual force would be proportional to just $\alpha\mu^2$, that is,

$$F_h - F_r = \tfrac{1}{2}\alpha\mu^2 \frac{G_0 M_r d}{(b^2+d^2)^{\frac{1}{2}}} - \alpha\mu^2 \frac{G_0 M_h a}{L}$$
$$\times\left[\ln 2 - \gamma + E_i(-\tfrac{1}{2}\mu L) - \frac{R_2^2 \ln \mu R_2 - R_1^2 \ln \mu R_1}{R_2^2 - R_1^2} + \dots\right]. \tag{7.87}$$

In the experiment of Spero *et al.*, the calibration ring was moved so that the test mass always lay on the ring axis at a fixed distance $(d = b/2^{\frac{1}{2}}$, where the force due to the ring was maximum and insensitive to the exact location of the test mass). Therefore, the detectable non-Newtonian force is given by

$$\delta F = \alpha\mu^2 \frac{G_0 M_h a}{L}\left[\ln 2 - \gamma + E_i(-\tfrac{1}{2}\mu L) - \frac{R_2^2 \ln \mu R_2 - R_1^2 \ln \mu R_1}{R_2^2 - R_1^2} + \dots\right]. \tag{7.88}$$

The expected relative change of the 'constant' G with distance is then

$$\frac{\Delta G}{G} = \tfrac{1}{4}\alpha\mu^2 L^2\left[\ln 2 - \gamma + E_i(-\tfrac{1}{2}\mu L) - \frac{R_2^2 \ln \mu R_2 - R_1^2 \ln \mu R_1}{R_2^2 - R_1^2} + \dots\right]. \tag{7.89}$$

In their experiment, Spero *et al.* looked for deflections of the balance as the cylinder was moved from side to side. The cylinder was made of non-

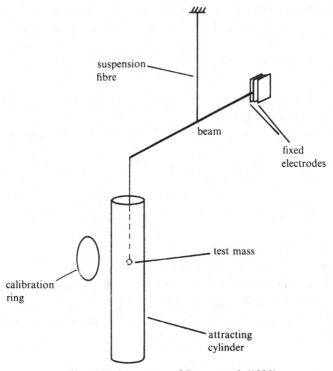

Fig. 7.22 Apparatus of Spero *et al.* (1980).

magnetic stainless steel 0.6 m long with a mass of 10.44 kg. The test mass, a cylinder of high-purity copper of mass 22 g and length 0.044 m was hung 0.83 m below the end of one arm of a torsion balance. The torsion balance arm was made of oxygen-free high-conductivity copper of an overall length of 0.6 m, suspended by a 75 μm diameter fibre, 0.32 m long (see Fig. 7.22).

The calibration ring was of copper of radius $b = 0.121$ m located so that the test mass was $b/2^{\frac{1}{2}} = 0.086$ m away from the surface of the ring on the axis. The mass was 133 g, and chosen so that the Newtonian force produced by the ring was approximately equal to the Newtonian residual force due to the hollow cylinder. In the calculation, the calibration ring was treated as a ring of negligible thickness since the outside and inside radii of the ring were respectively 0.1230 m and 0.1185 m, and the thickness was only 4.5 mm (Spero *et al.*, 1980).

The deflection of the torsion balance was measured with an optical lever, illuminated by an infrared-emitting diode. The reflected light fell upon a quadrant photocell, the signal from which was amplified in a differential amplifier. The torsion balance was in an enclosure evacuated to 4×10^{-6} Pa.

The hollow cylinder was placed with its axis vertical upon a trolley that could be moved automatically horizontally back and forth so that the test mass was alternately on the axis of the cylinder and near the inner surface, while at the same time the ring could be moved to cancel the residual Newtonian force. The recorded deflection of the torsion balance was examined for a signal of the same period as the motion of the cylinder. The test mass was also attracted by the trolley and other moving parts, and all such effects required careful calculation.

The quantity that was measured was the change in torque, ΔT_{exp}, on the torsion balance when the hollow cylinder was translated sideways by 34.3 mm. ΔT_{theo} is the theoretical change according to the inverse square law. It was found that

$$\delta = \Delta T_{\mathrm{exp}} - \Delta T_{\mathrm{theo}} = (0.002 \pm 0.14) \times 10^{-13} \text{ N m}. \tag{7.90}$$

In terms of the $\alpha\mu^2$ parameter,

$$\alpha\mu^2 < 0.029 \text{ m}^{-2} \tag{7.91}$$

or, if $\alpha = \frac{1}{3}$, as in the theory of Fujii (1972), the wavelength of the assumed particle would be

$$\mu^{-1} > 3.4 \text{ m}. \tag{7.92}$$

The result can also be expressed in an α–μ^{-1} plot as shown in Fig. 7.25. It is in obvious conflict with that of Long expressed in eqns (7.83) and (7.84) but agrees well with the results of Chen *et al.* (1984).

7.8 Second experiment of Newman and co-workers

So far we have discussed experiments carried out with the torsion balance at distances of the order of 0.1 m. The advantage of these experiments is that they can reach quite a high precision of measurement, say 10^{-4}. They were mainly undertaken to check Long's result at the same distance and they did indeed reach precisions of 1 part in 10^4. The purpose was limited and a closer test of the inverse square law requires much greater distances. The experiment of Hoskins *et al.* (1985) is one of the most precise experiments extending the test distance up to 1 m.

Fig. 7.23 shows the arrangement of the experiment. The torsion balance beam was a copper bar 0.6 m long and weighing 523 g suspended by a 90 μm diameter tungsten fibre 200 mm long attached to the mid-point of the beam. The copper bar was the test mass and was attracted by a cylindrical copper mass weighing 43 g placed 50 mm from one side of the beam and by two larger cylindrical copper masses weighing 7.3 kg carried on carts, each 1.05 m away from the axis of the torsion balance.

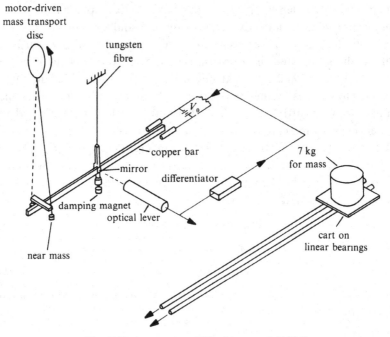

Fig. 7.23 Apparatus of Hoskins *et al.* (1985).

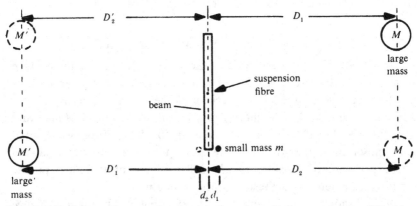

Fig. 7.24 Positions of attracted and attracting masses in the experiment of Hoskins *et al.* (1985). Alternative positions of masses shown by full and broken circles.

The masses of the small cylinder and two large cylinders and the relevant distances were chosen so that the total torque on the beam should be nearly zero; the residual torque was detected by observing the change of deflection when the attracting masses were simultaneously interchanged.

An important feature of the experiment is that the positions of the attracting masses are symmetrical with respect to the torsion arm so that the torque exerted is insensitive to first order to the position of the arm. Let the distances of the small mass from the torsion arm in the two positions be d_1 and d_2 and let the distances of the large masses M and M' be D_1 and D_1' and D_2 and D_2' respectively (see Fig. 7.24). At the balance condition, the relations between the torques according to the inverse square law are, in position 1,

$$m/d_1^2 = M/D_1^2 + M'/D_1'^2, \tag{7.93}$$

and in position 2,

$$m/d_2^2 = M/D_2^2 + M'/D_2'^2. \tag{7.94}$$

If the masses M and M' are nearly equal, and the distances D_1 and D_1' are also nearly equal, it can be shown that the error in setting

$$d_1 = [m/(M+M')]^{\frac{1}{2}}(D_1 + D_2') \tag{7.95}$$

and

$$d_2 = [m/(M+M')]^{\frac{1}{2}}(D_2 + D_1') \tag{7.96}$$

is of order $[(M-M')/M]^2$ and $[(D_1-D_1')/D_1]^2, [(D_2-D_2')/D_2]^2$ at most. Thus, to a close approximation,

$$d_1 + d_2 = [m/(M+M')]^{\frac{1}{2}}[(D_2'+D_1)+(D_1'+D_2)], \tag{7.97}$$

in which expression the individual values of D_1, D_2, D_1', D_2' and d_1, d_2 do not enter. Thus it is displacements of the masses, $(d_1 + d_2), (D_2' + D_1), (D_1' + D_2)$ that have to be measured and not their distances from the copper arm. To further ease the burden of metrology of the experiment, the small mass is located in the horizontal plane containing the balance bar, at that position along its length at which the torque produced is a maximum so that torque is insensitive to errors in the small mass position relative to the torsion balance.

The deflection of the torsion balance was measured by an optical lever: the light source was an infrared light emitting diode (LED), the receiver was a quadrant photodiode. The differential output of the photodetector was amplified, filtered, and recorded at 20 s intervals by a computer which not only monitored the motion of the beam but also controlled the experiment.

Damping of the pendulum is very essential for torsion balance experiments, as has been emphasized more than once, and in the experiment of Hoskins *et al.*, modes of oscillation other than the main torsion mode were damped by eddy currents induced in a copper cylinder fixed on the bottom of the balance by a stationary permanent magnet below. Eddy currents are most effective in damping the oscillations with

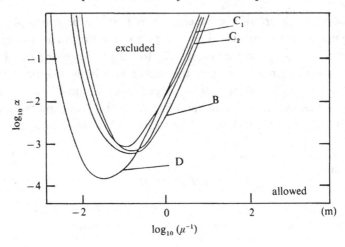

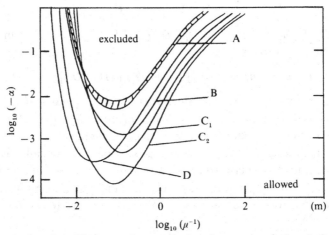

Fig. 7.25 Results of four experiments displayed on an α–μ^{-1} plot. A: Long (1976); B: Hoskins *et al.* (1985); C_1: Chen *et al.* (1984), null; C_2: Chen *et al.* (1984), non-null; D: Spero *et al.* (1980).

relative high speed and do not significantly damp the main torsional mode which is much slower than any others. The torsion mode was damped by velocity feedback on electrodes fixed near the torsion beam. The gain of the servo-loop was adjusted to obtain approximately critical damping.

The result of experiments over three days was that the ratio of the torque T_{105} exerted by the two masses at 1.05 m, to the torque T_5 exerted by the small mass at 0.05 m, was

$$R_{\text{exp}} = 1.018\,37 \pm 0.000\,37. \qquad (7.98)$$

The torques according to inverse square law attractions were computed for all parts relative to each other by an integral of six dimensions, extending over the volumes of the bar and all attractors, with the theoretical result

$$R_{\text{theo}} = 1.018\,25 \pm 0.000\,58. \tag{7.99}$$

Comparing the above results, the quantity $\Delta G/G$ is found to be

$$\Delta G/G = (1.2 \pm 7) \times 10^{-4}, \tag{7.100}$$

and the limiting value of the parameter $\alpha\mu^2$, calculated according to the non-Newtonian formula, is

$$\alpha\mu^2 < 0.0013. \tag{7.101}$$

That is the smallest value from any of the foregoing experiments.

The experimental results of Long (1976), Spero *et al.* (1980), Chen *et al.* (1984) and Hoskins *et al.* (1985) are compared in an α–μ^{-1} plot as shown in Fig. 7.25, in which it can be seen that the allowed region of Long's experiment is inconsistent with the other experiments.

7.9 Experiment of Panov and Frontov

Another Cavendish type experiment at longer distance was carried out by Panov & Frontov (1980).

A torsion balance was made of a fused quartz beam 0.4 m long suspended by a tungsten fibre of length 310 mm and diameter 30 μm, with test masses weighing 10 g fixed at the end of the beam. The whole assembly had a natural period of 910 s. The balance was sealed in a glass vacuum chamber and the temperature was thermostatically controlled to 30 °C ± 0.002 deg.

The motion of the balance was monitored by a capacitive displacement sensor with a resolution of 10^{-8} rad; the oscillatory motion was damped actively by velocity feedback.

Panov and Frontov have made tests of the inverse square law with this torsion balance at the longest distance ($\approx$ 10 m) so far used. Because the forces are taken much weaker, the attracting masses were made much larger: $M_0 = 0.2013$ kg, $M_1 = 56.66$ kg and $M_2 = 594.9$ kg in order to obtain gravitational attractions matching the sensitivity of the balance. A resonance method was used: the torsion pendulum was set into motion by displacing the attracting mass periodically with the same period as that of the torsion pendulum.

The inverse square law was tested by comparing the torques produced by attracting masses at distances of 0.4206 m (r_0), 2.958 m (r_1) and 9.84 m (r_2) from the test mass. The results are: $G(r_1)/G(r_0) = 1.003 \pm 0.006$ and $G(r_2)/G(r_0) = 0.998 \pm 0.013$ and their interpretation in terms of the limits of the product $\alpha\mu^2$ is given in Table 7.1.

7.10 Direct measurement of the gradient of gravitation – Paik's method

The experiments discussed above involve the direct measurement or comparison of the gravitational force. Paik (1979) proposed a new 'source-independent' null experiment to test the Newtonian gravitational law by means of measurements of the gradient of the gravitational force instead of the force, so that problems associated with absolute measurements in metrology could be avoided. A prototype experiment performed by Chan, Moody & Paik (1982) is described below.

A higher precision may be expected in the future.

Principle

While the gravitational force at points inside a spherical shell or inside an infinitely long hollow cylinder vanishes if the Newtonian law holds, there will be such a force if the Newtonian law is violated, and that is the basis of the null experiment of Spero *et al.* (1980). However, experiments that are the attraction of a local mass have an important shortcoming: deviations from perfect geometry, the uncertainties of the dimension and the inhomogeneities of the density of the attracting mass may place severe limitation on the precision. To see how serious that might be, we take the experiment of Spero *et al.* as an example. In their experiment using a hollow cylinder as an attracting mass, they paid great attention to possible errors caused by any non-uniformity of the cylinder. They measured the dimensions to 10^{-7} m, and checked the density distribution by γ-ray scanning, yet even so the error contribution (in terms of torque) was still estimated to be as high 6×10^{-15} Nm compared with the total error which was 14×10^{-15} Nm. Generally speaking, because of the limit of the metrology associated with measurements of separations, dimensions and density distributions, the precision of the experiments using null measurements may be limited to 1 part in 10^4.

The source-independent method proposed by Paik (1979) involves a measurement of the gradient of the gravitational attraction independent of the geometry of the attracting mass. If the inverse square law holds for an arbitrary distribution of attracting mass, the gradient of the force produced by this mass is only dependent on the local density $\rho(r)$, i.e. the potential satisfies Poisson's equation

$$\nabla^2 V(r) = -4\pi G\rho(r), \qquad (7.102)$$

where $\rho(r)$ is the actual density of matter at the point P (coordinate r), where the detector is located. Any deviation from the inverse square law will produce an extra term on the right side dependent on the forms of non-local masses. Thus, Poisson's equation of eqn (7.102) will be replaced by

$$\nabla^2 V(r) = -4\pi G\rho(r)[1 + \alpha(1 + \mu r)\,e^{-\mu r}] - \alpha\mu^2 \int_u \frac{G\rho(r)}{r^2} r\,e^{-\mu r}\,du, \quad (7.103)$$

where u is the volume of the attracting mass. The first term in eqn (7.103), like the Newtonian analogue, depends only on the local density at the field. The second term is proportional to the total non-Newtonian potential at the field point $P(r)$. Thus, Paik proposed measuring the Laplacian of the potential near a gravitational source and comparing it with zero.

Although there may be many ways to make such a measurement, one practical way is to use a gravity gradiometer to measure the gradients of the attraction along three orthogonal directions because, in a Cartesian coordinate system, the Laplacian can be written as

$$\nabla^2 V = \frac{\partial^2 V}{\partial x^2} + \frac{\partial^2 V}{\partial y^2} + \frac{\partial^2 V}{\partial z^2} = \frac{\partial g_x}{\partial x} + \frac{\partial g_y}{\partial y} + \frac{\partial g_z}{\partial z}, \qquad (7.104)$$

where g_x, g_y and g_z are the gravitational forces along x, y and z directions, that is, the Laplacian is the sum of the gradients of the three components of the gravitational force parallel to three orthogonal directions. To avoid the extraneous disturbances caused by the d.c. background, it is desirable that the force of a local source should be modulated in amplitude at a definite frequency, for then a modulation technique can be used to reduce the background noise.

Preliminary experiment

Based on the above principle Chan *et al.* (1982) carried out an experiment to test the inverse law. The gravitational force was produced by a local mass of 1.6 tonne. To have an intensity-modulated attraction, the mass was hung by a wire to form a simple pendulum with a frequency of 0.228 Hz. The initial amplitude of the swing was 0.90 m and over a time period of 2000 s (which is the measurement period) the amplitude decayed by 17 per cent. The gradiometer was located 2.3 m away from the

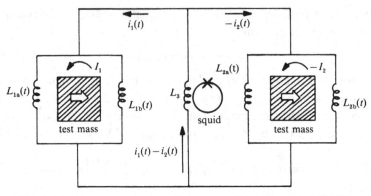

Fig. 7.26 Superconducting gradiometer used by Chan *et al.* (1982).

pendulum, and so the range of distance for the test of the inverse square law was 1.9–2.7 m. A single-axis superconducting gravity gradiometer was used and rotated in three orthogonal orientations to measure the gradients of the gravitational attraction.

Anandan (1984a, b) showed that a gravitational acceleration can be detected by a change of current in a superconducting circuit containing two superconducting solenoids with a steady current in the loop. A practical superconducting accelerometer was first proposed and developed by Paik and Fairbank (see Paik, 1976) by using a niobium proof mass (test mass) and an rf SQUID (superconducting quantum interference device), a Josephson junction magnetometer.

The superconducting gravitational gradiometer used in the experiment of Chan *et al.* is the combination of two superconducting accelerometers in a symmetrical differentiating mode. Fig. 7.26 gives the schematic diagram of such a gradiometer, which is known as a 'current differencing' gravitational gradiometer.

In Fig. 7.26, L_{1a}, L_{1b} and L_{2a}, L_{2b} are the two sensing coils for the test masses M_1 and M_2 respectively. M_1 and M_2 are separated by a distance l in order to measure the difference of the accelerations acting on them.

The two pairs of sensing coils have loop superconducting currents I_1 and I_2 respectively and they share one SQUID with the input coil L_3. Because of the 'Meissner effect', the magnetic field produced by I_1 and I_2 cannot penetrate the niobium diaphragm.

In a non-uniform gravitational field, M_1 and M_2 have slightly different accelerations along the line of l. The difference of the motions of the test masses due to the external gravitational field will vary the magnetic fluxes passing through the two pairs of sensing coils, thus inducing a signal $i_1 - i_2$

in L_3, the input solenoid of the rf SQUID. The ratio I_1/I_2 is then adjusted so that $i_1 - i_2$ is proportional to the difference of the accelerations acting on the test masses M_1 and M_2, that is, to the gradient of gravity along the line joining the two masses.

The difference of acceleration, which has a unit of s^{-2}, represents the gradient along a certain direction. In the experiment of Chan *et al.* (1982), the test masses M_1 and M_2 were 0.4 kg and l was 0.15 m. The natural frequency of the mass was 25 Hz. The whole device was housed in a vacuum and isolated by a pendulum suspension from a tower on the top of the cryostat to decouple it from the vibration of the floor. The instrument noise level was $10^{-10} \, s^{-2} \, Hz^{-\frac{1}{2}}$.

To measure $\nabla^2 V(r)$, the cryostat was turned incrementally by $120°$ on a turntable to rotate the gradiometer axis (M_1 to M_2) into three orthogonal orientations. The data were collected from the output of the SQUID and averaged over 500 swings of the lead pendulum. The measured quantities corresponding to eqn (7.104) were $10.07 \times 10^{-9} \, s^{-2}$, $-5.88 \times 10^{-9} \, s^{-2}$ and $-3.47 \times 10^{-9} \, s^{-2}$ respectively. After some necessary corrections, the final result was

$$\nabla^2 V = (0.15 \pm 0.23) \times 10^{-9} \, s^{-2}. \qquad (7.105)$$

Seismic noise was the main source of random error (as a matter of fact, another application of the device is to the measurement of the free elastic oscillations of the Earth), and there was a further significant error arising from errors in the condition of orthogonality.

From Table 7.1, it is seen that the precision of Paik's method is quite comparable to precise torsion pendulum experiments. Given that the experiment was only a preliminary one, there are clear possibilities of achieving a much better precision in the future by improving the geometry of the experiment with a simultaneous three-axis measurement; reducing the effect of seismic noise with a better isolation or a longer integration time; eliminating the effect of magnetic contamination with better shielding; increasing the testing distance with a heavier mass and larger swing range. It should be quite possible to extend Paik's method to test the inverse square law at more than 10 m.

7.11 Inverse square law at high frequency – Weber, Sinsky and Hirakawa method

All the tests of the inverse square law so far discussed were performed at a low frequency, the gravitational fields were constant or varied at a very low frequency, but as a by-product of the development of gravitational wave detectors, the validity of the inverse square law at frequencies from a few tens of hertz to a few kilohertz has been tested by Sinsky & Weber (1967), Hirakawa, Tsubono & Oide (1980) and Ogawa, Tsubono & Hirakawa (1982).

An experiment to test the inverse square law at a relatively high frequency needs a generator, the source of the alternating gravitational signal, and an antenna, the mechanical receiver to respond to the alternating signal. The alternating component of a gravitational attraction in the laboratory is superposed on the average component and is weaker than it, so that consequently, in all experiments, a resonant detector has been used, tuned to the alternating frequency.

The antenna is made of a large mass which may have various shapes (in the experiment of Sinsky and Weber, it was a cylindrical bar, in the experiments of Hirakawa and Ogawa, it was a rectangular plate) and different means of support (the bar was supported by a suspension wire at the centre where a nodal point of the fundamental oscillation was located, the rectangular plate was supported at four nodal points of the second harmonic vibration). The design of antennas corresponds to the form of the generator and the type of signal it radiated.

Experiment of Sinsky and Weber

This experiment was originally performed for the calibration of Weber's gravitational wave detector. It can also be interpreted as a verification of the inverse square law at a few kilohertz.

The antenna is an aluminium cylindrical bar 1.54 m long and 0.6 m in diameter, the generator is an aluminium cylindrical bar with nearly the same length and a diameter of 200 mm, as illustrated in Fig. 7.27.

The bar oscillates at its lowest frequency in the fundamental longitudinal mode. If damping is ignored, the longitudinal displacement of the bar, $\xi(z, t)$, at the axial position z and t satisfies the wave equation

$$\rho \frac{\partial^2 \xi}{\partial t^2} - Y \frac{\partial^2 \xi}{\partial z^2} = 0, \tag{7.106}$$

where ρ is the density of the cylinder and Y is Young's modulus for the material. The speed of sound in the material is

$$v_s = \left(\frac{Y}{\rho}\right)^{\frac{1}{2}}. \tag{7.107}$$

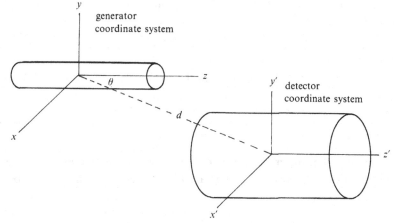

Fig. 7.27 Experiment of Sinsky and Weber (1967).

With the centre of the bar rigidly clamped and the two ends free, the appropriate solution of eqn (7.106) is

$$\xi(z, t) = \sum_{n=1}^{\infty} A_n e^{i\omega_n t} \cos k_n z, \qquad (7.108)$$

where A_n is the amplitude for the nth mode, and

$$\omega_n = \frac{n\pi v_s}{L}, \qquad (7.109)$$

$$k_n = \frac{n\pi}{L} \qquad (7.110)$$

(L is the length of the antenna). The generator was driven at the fundamental frequency $f_1 = v_s/2L$ by two large piezoelectric transducers bonded on the surface near the centre; the amplitude of the strain was 5×10^{-5}. The longitudinal Newtonian attraction between the two cylinders along the axial (longitudinal) direction of the cylinders can be calculated using the cylindrical formula in Chapter 6 if the inverse square law is valid. Because of the periodic distortion of the generator, the attraction will be modulated correspondingly. The generator affects the antenna in two ways: by the motion of the particles in the generator described in eqn (7.108) and also by the change in the density because of the periodic change of the length of the cylinder.

The length of the generator at time t in the fundamental oscillation is

$$L = L_0\left(1+\frac{A_1}{L_0}\sin\omega_1 t\right), \tag{7.111}$$

where A_1 can be expressed in terms of the strain ε_0 at the centre of the bar, as

$$A_1 = \frac{L_0}{\pi}\varepsilon_0. \tag{7.112}$$

If we assume that the cross-section of the cylinder remains constant during the oscillation of the cylinder, the variation of density with time will be

$$\rho-\rho_0 = -\rho_0\frac{\varepsilon_0}{\pi}\sin\omega_1 t. \tag{7.113}$$

The Newtonian attraction between a particle in the antenna located at $(x,0,z)$ and the whole generator is then

$$a = 2G\rho_0\left\{C_t[z+\tfrac{1}{2}L_0, R, x]-C_t\left[z-\tfrac{1}{2}L_0\left(1+\frac{A_1}{L_0}\sin\omega_1 t\right), R, x\right]\right\}$$
$$-2G\rho_0\frac{A_1}{L_0}\sin\omega_1 t\{C_t[z+\tfrac{1}{2}L_0, R, x]-C_t[z-\tfrac{1}{2}L_0, R, x]\}. \tag{7.114}$$

If we subtract the stationary component of the attraction which is

$$a_{st} = 2G\rho_0\{C_t[z+\tfrac{1}{2}L_0, R, x]-C_t[z-\tfrac{1}{2}L_0, R, x]\}, \tag{7.115}$$

the variable attraction, according to the inverse square law, is then

$$a_{dyn} = 2G\rho_0\left\{C_t[z-\tfrac{1}{2}L_0, R, x]-C_t\left[z-\tfrac{1}{2}L_0\left(1+\frac{A_1}{L_0}\sin\omega_1 t\right), R, x\right]\right\}$$
$$-2G\rho_0\frac{A_1}{L_0}\sin\omega_1 t\{C_t[z+\tfrac{1}{2}L_0, R, x]-C_t[z-\tfrac{1}{2}L_0, R, x]\}. \tag{7.116}$$

Thus, the displacement $\xi'(z', t)$ of an antenna which has its fundamental longitudinal oscillation frequency close to that of the generator and is near enough to it for the time to be the same for both cylinders, satisfies the equation

$$\rho_a\frac{\partial^2\xi'}{\partial t^2} - Y_a\frac{\partial^2\xi'}{\partial z^2} - \gamma_a\frac{d\xi'}{dt} = \rho_a a_{dyn}(x, z), \tag{7.117}$$

where ρ_a, Y_a, γ_a are respectively the density, Young's modulus and the damping factor for the material of the antenna. The solution of the above equation was calculated numerically for the following values: $\rho_a = \rho_0 = 2.78 \times 10^3$ kg m^{-3}, $Y_a = Y_0 = 7 \times 10^{10}$ N m^{-2}, $\omega_1 = 10^{-4}$ s, $\varepsilon_0 = 5 \times 10^{-5}$. The gravitational inverse square law can be tested when the generator and the antenna are located at different distances. In the experiment of Sinsky and Weber, the generator and the antenna were placed in four different positions:

(A) The two cylinders were located coaxially with the distance between the two centres equal to 1.72 m.
(B) The two cylinders were located coaxially with the distance between the two centres equal to 1.84 m.
(C) The two cylinders were placed parallel with the axes 0.2 m apart and the distance between the centres equal to 1.72 m.
(D) The two cylinders were placed parallel with the axes 0.3 m apart and the distance between the centres equal to 1.72 m.

The strains were measured by quartz piezoelectric crystals bonded to the surface of the antenna. In order to convert the output voltage of the detectors into the actual strain of the antenna, a lengthy calculation involving the characteristics of the cylinder, the sensing crystals, the associated electric circuits and the amplifiers had to be made. The final result was expressed by P, the induced power increase of the antenna by the generator.

Sources of disturbance, acoustic, electromagnetic and vibrational, were carefully eliminated. The generator and the antenna were housed in vacuum chambers carefully decoupled from the ground by vibration isolators consisting of alternate layers of felt, steel and rubber.

Each set of measurements lasted 6 hours during which 360 voltages were recorded.

To interpret the experimental results as a test of the inverse square law, we have,

$$\frac{\Delta G}{G} = \frac{\varepsilon_{exp} - \varepsilon_{theo}}{\varepsilon_{theo}} = \frac{\Delta \varepsilon}{\varepsilon_{theo}}, \tag{7.118}$$

where ε_{exp} and ε_{theo} are the measured and calculated strain of the antenna. As we mentioned the results were given as a relative increase of power, which is proportional to the square of the included strain, and so in terms of the observed quantities

$$\frac{\Delta G}{G} = \frac{1}{2} \frac{\Delta P}{P}. \tag{7.119}$$

Table 7.7. *Results of Sinsky and Weber*

Group	Distance (m)	Power Measured (%)	Power Calculated (%)	$\alpha\mu^2$ (m^{-2})	Sign of α
A	1.72	21.0 ± 2.1	21.7 ± 2.0	< 0.5	$-$
B	1.84	8.3 ± 1.7	7.6 ± 0.7	< 0.1	$+$
C	1.72	14.5 ± 2.1	13.2 ± 1.2	< 0.1	$+$
D	1.72	7.9 ± 2.6	6.5 ± 0.6	< 0.2	$+$

Sinsky & Weber (1967).

Then, eqn (7.20) can be used to estimate the deviation $\alpha\mu^2$ from the Newtonian law and the factor $\tilde{R}^2$ would be represented by the square of the characteristic distance d^2 for simplicity.

Table 7.7 gives the results of the four groups at different distances. From these results, it can be seen that the inverse square law is satisfied within the experimental accuracy. Although limits of $\alpha\mu^2$ are not as tight as in other experiments listed in Table 7.1, they constitute the first test of the Newtonian law in the kilohertz frequency range.

The major advantage of the method of Sinsky and Weber is the high sensitivity. Nowadays, the sensitivity of strain measurements can reach 10^{-21} or less and so provide great possibilities of extending the interaction distance up to 10 m. Hirakawa *et al.*, in their later development, exploited those possibilities.

Experiments of Hirakawa and Ogawa

The experiment of Sinsky and Weber could be improved in two ways. First, the strain of the generator was only about 5×10^{-5}, clearly inadequate but limited by the method used to drive the bar and by the yield point of the material used. Secondly, the bandwidth of the antenna (of order 0.01 Hz), although it reduces the noise, makes the coupling difficult. The natural frequencies of the generator and antenna are decided by the lengths of the cylinders, and various disturbances, particularly a change of temperature, can change the lengths of the cylinders and destroy the resonance condition. In the experiment of Sinsky and Weber, the frequency of the generator was kept in tune with that of the antenna by a heating coil

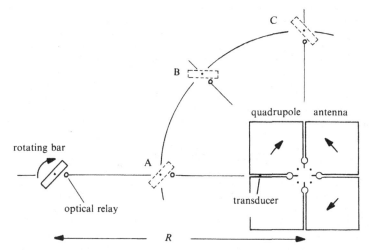

Fig. 7.28 Experiments of Hirakawa *et al.* (1980) and Ogawa *et al.* (1982).

attached to the surface of the cylinder of the generator. Both problems were dealt with in the experiments of Hirakawa and Ogawa and better precision was attained.

In the experiments of Hirakawa *et al.* (1980) and Ogawa *et al.* (1982) (Fig. 7.28), the antenna was a square aluminium plate (1.65 m × 1.65 m × 0.19 m) of mass 1400 kg with cuts on each side. It formed a quadrupole antenna and was supported on its four nodal points, so that the resonant frequency, 60.8 Hz, was twice the fundamental frequency. Oscillations of the antenna were sensed by an electric transducer mounted inside one of the cuts, as indicated in Fig. 7.28. The generator was a steel bar rotated at a frequency of 30.4 Hz. In the early version of the experiment Hirakawa *et al.* (1980) employed a bar 0.52 m long with a mass 44 kg for the short distance test, while in the later version of the experiment, Ogawa *et al.* (1982) used a bar 0.9 m long and of mass 401 kg for tests at longer distances up to 10 m.

As in the experiment of Sinsky and Weber, the natural bandwidth of the antenna was too narrow for proper coupling. The mechanical quality factor Q for the square plate was 5.5×10^4, and so the bandwidth was 10^{-3} Hz compared to the frequency stability which was about 10^{-3} Hz h^{-1}.

To overcome this difficulty, electrical damping was used to increase the bandwidth. Electrical damping, as discussed in Chapter 3, can effectively reduce the mechanical quality factor of a linear oscillator without introducing extra noise and can be cooled by part of the feedback system to reduce the effect of thermal noise without having to cool the whole

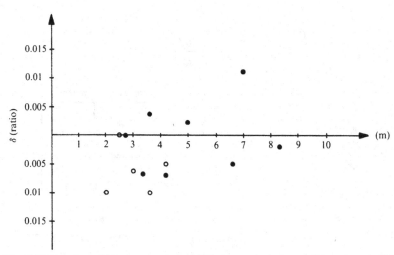

Fig. 7.29 Results of Hirakawa *et al.* (1980) (open circles) and Ogawa *et al.* (1982) (solid circles).

apparatus. The cooling resistor in the experiment was cooled to a temperature 1.78 K. The resulting mechanical quality factor was reduced to 5.3×10^3, the effective temperature T_e was 52 K. The effective bandwidth had been broadened to 0.01 Hz.

It was extremely important to avoid non-gravitational coupling of the generator and the antenna. The rotating bar was housed in a steel vacuum tank which was isolated from the laboratory floor by four pneumatic springs and surrounded by a 3 mm thick aluminium soundproof wall. Any magnetic effect at the frequency of 60.8 Hz was cancelled by a pair of Helmholtz coils placed around the tank.

The phase noise of the rotor may be an important source of noise. Consequently, the phase of the motor that drove the rotor was locked to a quartz frequency synthesizer and was stable to within about 10^{-3} rad over a period of 2 hours.

Because the generator was made of a rotating bar, the gravitational signal it produced was much stronger than in the experiment of Sinsky and Weber, and so the distance between the generator and antenna could be quite considerable. Fig. 7.29 shows the results of the experiment of Hirakawa *et al.* (1980) with distances from 2.2 m to 4.2 m, and those of Ogawa *et al.* (1982) with distances from 2.6 m to 10.7 m; the amplitude of the antenna is shown as a function of distance.

The result of the first experiment confirmed the inverse square law within the experimental range of distance with an accuracy of ± 3 per cent, giving $\delta = (0 \pm 5.3) \times 10^{-2}$ for a potential of the form $1/R^{1+\delta}$; the second experiment gave $\delta = (2.1 \pm 6.2) \times 10^{-3}$. The limiting value for the parameter $\alpha \mu^2$ is

$$\alpha \mu^2 < 0.03 \text{ m}^{-2}. \tag{7.120}$$

8

The constant of gravitation

8.1 Introduction

The gravitational constant G is defined by Newton's law as

$$F = G\frac{M_1 M_2}{R^3}R, \tag{8.1}$$

where M_1, M_2 are the passive and active gravitational masses of two point bodies respectively, F is the force of attraction and R is the distance between them. Although G was the first physical constant to be introduced and measured in the history of science, it is still the least well known.

It was 100 years ago that Boys (1889) read his paper before the Royal Society of the London, claiming that his apparatus would be able to determine G to a precision of 1 part in 10^4; this goal has never been achieved either by himself or by others after him. Table 8.1 lists the main experimental results over the past two decades. Although most of the individual experiments in the table were stated to have uncertainties of about 10^{-4}, the disagreement between them is near 1 part in 10^3. The generally accepted value, the current CODATA value (Cohen & Taylor, 1986), is

$$G = (6.672\,59 \pm 0.000\,85) \times 10^{-11} \text{ m}^3 \text{ s}^{-2} \text{ kg}^{-1}.$$

An accurate knowledge of G is not only important from the point of view of theoretical physics, but also significant for practical purposes, particularly when finding the density and density distributions of the interiors of the Earth, Moon, planets and stars. For this reason, great efforts have been made for many centuries to obtain a reliable value. There are about a thousand publications in the literature concerning the determination of G; many famous physicists have devoted a great deal of work to it.

196

Table 8.1. *Recent published figures for the gravitational constant*

Authors	Year	G $(10^{-11} \text{ m}^3 \text{ kg}^{-1} \text{ s}^{-2})$
Rose *et al.*	1969	6.6699 ± 0.0014
Facy & Pontikis	1970, 1971	6.6714 ± 0.0006
Renner	1974	6.6700 ± 0.008
Karagioz *et al.*	1976	6.668 ± 0.002
Sagitov *et al.*	1978	6.6745 ± 0.0008
Luther & Towler	1982	6.6726 ± 0.0005
de Boer	1987	6.667 ± 0.0007

See also Sagitov (1976).

Earlier, Poynting (1894) and Mackenzie (1900) summarized the contemporary knowledge of G in 1894 and 1900 respectively. Recently, de Boer (1984) made a survey of recent major experiments and Gillies (1987) published a very comprehensive report giving an index of the measurements of G, containing over 1200 references.

The reason that, in spite of the considerable effort, G is still not very well known, is not only the weakness of gravitational attraction on a laboratory scale, but also the demands of absolute measurements. Although in the experiments discussed in the Chapter 4 to test the weak principle of equivalence, the precision is limited by the weakness of the gravitational attraction, the measurements are relative and no absolute measurement is involved, as a result the precision of tests can reach 10^{-11} or even better. In the determination of G, however, the situation is different. G is defined by three fundamental quantities – time, length and mass. The measurement of G requires that absolute values be measured for the masses of the test bodies and the attracting bodies, the separation, the period of the motion of the mechanical oscillator and so on, all of which may give rise to considerable experimental difficulties.

Newton himself estimated that measurements of G in the laboratory were impracticable but thought it could be estimated from attractions of mountains. He once made an ingenious estimate of the mean density of the Earth which was five or six times that of water. Then he considered two ways of verifying that estimation. First, the pull of a natural body (e.g. a mountain) was compared with a laboratory body (e.g. mass of a plumb-bob). The second was to compare the pull of the Earth with a laboratory sphere. Newton's idea has further developed into two widely used geological methods for the measurement of G. The first experimental estimates of G were in fact obtained from the attraction of a large local natural mass or of the Earth itself (see Maskelyne, 1775; Carlini, 1824; Airy, 1855). To see the advantage of the laboratory measurement of G, a brief review of measurements using a natural mass and the Earth is given below.

Use of a natural mass

Usually a mountain with a regular shape is chosen as such a mass. The gravitational attraction of the mountain can be measured in two ways.

(1) A pendulum is used to determine the local gravity g at different positions, first at the bottom of the mountain, then at the top. Since the mountain is irregular, the determination of G from g would require very complicated calculations. However, if the mountain can be treated as a perfect sphere of mass M, then G can be calculated from the following formula in its simplest form:

$$G = \frac{R^2 l}{M}\left[\left(\frac{2\pi}{T_2}\right)^2 - \left(\frac{2\pi}{T_1}\right)^2\right], \tag{8.2}$$

where R denotes the distance from the pendulum to the centre of the mountain when the pendulum is at the top, T_1 and T_2 refer to the periods at the first and second positions respectively and l is the length of the pendulum.

The method is limited by two factors. The first is the irregular shape of the mountain, and the second is the estimation of the density distribution from which the total mass of the mountain is to be calculated.

(2) A plumb-line is used to measure the deflection caused by the gravitational attraction of the local natural mass. In this direct measurement, the plumb-line is located in different places relative to a mountain. Different deflections are recorded, from which G is calculated. In the earliest experiments, fixed stars were used as references to measure the deflection.

The early measurements of G and the mean density of the Earth were made possible by the methods illustrated above. In the first published work on G, Bouguer (1754) used both methods, pendulum and plumb-line. Although in his work no precision could be expected, he did prove that the methods were promising. They were subsequently used by Maskelyne (1775, 'G' = 7.8), Carlini (1824, 'G' = 7.43), and others. In modern times,

Ni (1980) and his students used the same method to test the inverse square law of gravitation, where the local natural mass was replaced by a large oil tank.

Use of the Earth

This method was proposed by Airy (1856) who was the first to compare gravity (g) at the top and bottom of a mine by measuring the period of a pendulum. The ratio of the pull of the whole Earth to the pull of the Earth minus a certain thickness of the shell is obtained. Then the gravitational constant can be written in terms of g_1 and g_2 at different places of the mine, the depth l of the mine, the radius R and the mass M of the enclosed Earth and the latitude ϕ of the location of the experiment. First let us consider the simplest case; if the Earth is assumed to be made up of homogeneous and concentric shells, gravity can be expressed approximately as

$$g = \frac{GM}{R^2} - \omega^2 R \cos^2 \phi, \tag{8.3a}$$

and so

$$g_1 - g_2 = \left(\frac{2l}{R}\right)\frac{GM}{R^2} - \omega^2 l \cos^2 \phi \approx \frac{2l}{R}g_1 - \omega^2 l \cos^2 \phi. \tag{8.3b}$$

From the above expression, we can see that the quantities g_1, g_2, l, ω, R and M can be measured quite precisely. If the mass of the outer shell can be estimated reliably, an accurate value of G can be obtained. Airy found $G = 6.67$.

The above calculation is made under the assumption that the Earth has a spherically symmetric layered structure. To correct for the actual ellipsoidal shape of the Earth, it is assumed that the Earth has an ellipsoidal surface with constant density and the same ellipticity. It is debatable whether this assumption is right. Nevertheless, by accepting that possible systematic errors impose 2×10^{-2} uncertainty in the measurement, Stacey & Tuck (1981) obtained a figure for the value of G equal to 6.7 ± 0.13.

Michell was the first person to build an apparatus to measure G in the laboratory but, unfortunately, did not live to work on the experiment. His apparatus was given by Wollaston of the University of Cambridge to Henry Cavendish, who continued the work, made many improvements, and performed the first measurement of G in the laboratory (Cavendish, 1798).

Since then, more than 20 laboratory determinations of G have been made. Most are of historical interest only and few attained better results than Michell. We do not say much about the history nor describe them in detail except for explaining the principles and any noteworthy techniques involved.

The first method to be used (by Michell) is that of direct deflection – the gravitational force upon a test mass causes the displacement of the equilibrium position of a mechanical oscillator such as a torsion balance or beam balance. We consider first the use of a beam balance.

8.2 Direct deflection measurement with beam balance

The beam balance is a device commonly used for mass comparisons. With a properly designed beam balance, a resolution of 10^{-10} m s^{-2} can be achieved (see Speake & Gillies, 1987). Since there is no difficulty in the laboratory in producing a gravitational attraction of 10^{-6} m s^{-2}, a measurement of G with a precision of 10^{-4} may in principle be expected.

Principle

The principle of using a beam balance to measure the gravitational constant is shown in Fig. 8.1. Suspended from two ends of a beam balance, attracted masses a and b are first balanced against each other when the attracting mass M is absent, as shown in Fig. 8.1(a). Then a large mass M is moved near one of the attracted masses, mass b, say. The gravitational attraction between mass M and the test mass b will deflect the beam balance through an angle θ, as shown in Fig. 8.1(b).

When the beam is deflected by the attraction of mass M, there is a restoring force which is proportional to the deflection and which can then be calculated from the angle of deflection and the balance constant. They can be derived from the mechanical parameters of the system (the periodic time, the moment of inertia).

However, to take the full advantage of the balance, the torque produced by the attracting mass M should be balanced by another known force F acting at certain distance L from the fulcrum of the beam to return the beam to the original equilibrium position, as shown in Fig. 8.1(c). The gravitational torque is then obtained in terms of the applied torque. Naturally, the calibration of the applied torque is very important.

A practical problem immediately presents itself. Because the resolution of a beam balance is limited, a heavy attracting mass should be used to

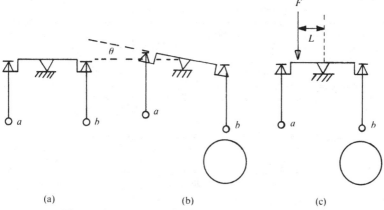

Fig. 8.1 Measurement of *G* with a beam balance.

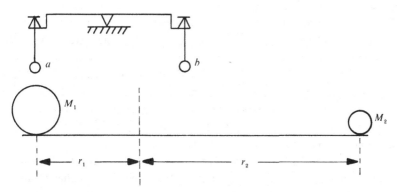

Fig. 8.2 Elimination of effect of tilt of floor on beam balance.

produce an adequate gravitational attraction, but moving a heavy mass will cause serious local tilting of the ground in the same direction as the deflection of the beam caused by the gravitational attraction.

Two configurations can be used to avoid that local tilt caused by the motion of the heavy attracting mass.

In the first (Fig. 8.2), two masses, one much heavier than the other, are placed on a rotatable table. The gravitational attraction on the test mass is greatest when the heavier mass M_1 is directly below the mass *a*, but there will be no change of tilt when the table is rotated provided the distances r_1 and r_2 satisfy the lever condition

$$M_1 r_1 = M_2 r_2.$$

That configuration was adopted by Poynting (1891) and by Speake &

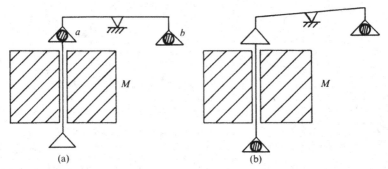

Fig. 8.3 Alternative configuration for elimination of effect of tilt of floor on beam balance.

Gillies (1987) in their determinations. Speake and Gillies, however, for convenience of arrangement, placed the attracting masses above the beam balance instead of below it.

The second configuration involves the replacement of the attracted mass instead of the attracting mass and is only used when the attracting mass is too heavy to move.

An auxiliary pan (see Fig. 8.3(a)) is added to one of the ends of the beam with two usual pans balanced against each other. The added pan is suspended with a long wire below the attracting mass. When the attracted mass is taken from the top pan to the bottom one as shown in Fig. 8.3(b) the deflection of the balance will change.

This arrangement was adopted by von Jolly (1881) and by Richarz & Krigar-Menzel (1898), except that the latter used two auxiliary pans to exchange the attracted masses from side to side, so doubling the deflection.

The advantage of this configuration is that it allows the use of a very large attracting mass in the laboratory. Richarz & Krigar-Menzel (1898) used an attracting mass of 100 tonnes constructed of many individual small blocks of lead.

Sensitivity of beam balance

The sensitivity of a beam balance, like that of a simple pendulum or torsion balance is determined by the smallest acceleration that can be detected by the balance regardless of noise; it is determined by the mechanical characteristics of the device and its associated detection system.

The mechanical behaviour of a beam balance is very complex. Metherell & Speake (1983) analysed the mechanical motion of a double pan balance with zero radius knife-edges as central and end fulcrums. The beam balance may be thought of as a compound pendulum rotating about the central fulcrum and two simple pendulums with two pans and their masses as pendulum bobs swinging about the two end fulcrums respectively. Metherell and Speake found that the motion of the beam balance is the superposition of three normal modes with three distinct eigen-frequencies, which are determined by the distance between the centre of gravity and the central pivot axis, the masses of the pans and the test mass, and the moments of inertia. However, among the three eigen-frequencies, the frequency ω_c of the 'compound pendulum' is the fundamental one. To avoid the complication of a full derivation, that frequency can be used to estimate the sensitivity of the beam balance. The acceleration to which the balance is subject is thus given by

$$a = l\theta\omega_c^2 = l\theta\frac{Mhg}{I}, \qquad (8.4)$$

where h is the distance between the pivot point and the centre of gravity of the beam balance including two masses and two pans, I and M are the moment of inertia and the mass of the beam balance including two test masses and two pans, l is the half-length of the beam arm, θ is the deflection angle and g is the local acceleration due to gravity. The least measurable acceleration is

$$\Delta a = \alpha g\Delta\theta, \qquad (8.5)$$

where $\Delta\theta$ is the resolution of the detector system and

$$\alpha = \frac{Mhl}{I}; \qquad (8.6)$$

α is a dimensionless coefficient, and its value is always less than 1.

By means of that formula we may compare the sensitivity of a beam balance with that of a simple pendulum or a torsion balance. The sensitivity of a beam balance is higher than that of a simple pendulum by the factor α^{-1} (see eqn (4.19)), while it may attain nearly the same resolution as the torsion balance (see eqn. (4.61)) if the height h is small enough.

From eqns (8.4) and (8.5), we can see that there are thus two ways of improving the sensitivity of a beam balance.

First, the value of α may be reduced by making h as small as possible. The centre of gravity of the whole balance must be adjusted to be as close as possible to the central pivot without causing instability. It can be adjusted to within a few micrometres; the fundamental period of the balance is then correspondingly very long.

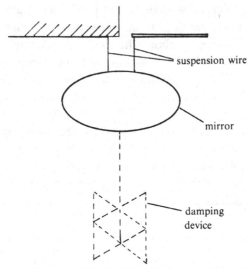

Fig. 8.4 Magnification of deflection of mirror with double suspension.

Secondly, the angular resolution can be improved and great efforts have been made to do so. Using a differential capacitive transducer, Speake and Gillies obtained a resolution of 10^{-8} rad. In the classical experiments of Poynting (1891) the angular resolution was improved by a 'double suspension' mirror (see Fig. 8.4). The deflection of the beam may be detected by an optical lever, but the double suspension mirror can increase the deflection to many times that of the simple optical lever. The double suspension mirror is a mirror suspended by two thin parallel wires, one from a fixed point, the other from a point attached to the beam of the balance. The distance d between two wires should be kept small (a few millimetres), since the angle through which the mirror turns for a given motion of the moving point on the beam is inversely proportional to this distance. By making d very small compared to the distance L of the moving point from the pivot of the beam balance, the angular magnification, which is L/d, can be made very large. It was 150 in the experiments of Poynting (1891).

To apply this mirror system to the balance, the mirror should be as light as possible and its oscillations should be easily damped. In Fig. 8.4 the broken line shows a simple damping device adopted by Poynting: with four light copper vanes attached to the end of the mirror, air damping was very effective.

With modern technology, it is not difficult to reach an angular resolution of 10^{-8} rad or better, although the calibration procedure may be rather complicated. To take an example, if $\alpha = 0.00001$, $\Delta\theta = 10^{-8}$ rad, and according to eqn (8.5) the sensitivity of the beam may reach 10^{-12} m s^{-2}, an impressive sensitivity.

Design of beam balance experiment

The beam balance has distinct advantages in the determination of G. They are as follows.

(1) It enables a very large attracting mass to be used, always an advantage given that gravitational forces are weak. Further, the larger the mass, the more it behaves as a point mass. A typical dimension R of an object of given density is proportional to $M^{\frac{1}{3}}$ while the distance D, at which the object produces a given force, is proportional to $M^{\frac{1}{2}}$. Thus the ratio R/D is proportional to $M^{-\frac{1}{6}}$ and approaches the value for a point mass, zero, more closely the greater M is.

Since the attraction of a large mass at a great distance is closer to that of a point mass, the geometrical form and homogeneity of density of a large mass at a great distance are less important than for a small mass at a shorter distance.

(2) The attraction on the beam balance is vertical and so effects of local extraneous gravitational attraction are much reduced as compared with their influence on a torsion balance. Moving bodies near the beam balance, provided they do not cause vibration of the ground or tilt of the floor and are not too close to the apparatus, will not disturb the balance.

As against its advantages, the beam balance has significant limitations as compared with a torsion balance.

(1) *Thermal noise.* Thermal noise could be the limiting factor for the performance of a highly sensitive beam balance because the Q of a beam balance is usually much lower than that of a torsion balance. It is also found (Speake & Gillies, 1987) that the value of Q may be less for larger amplitudes of swing. The Q of a normally designed beam balance will usually not be more than 100.

Further, although the Q may be already quite low, beam balances are usually damped electromagnetically to reduce the time of observation. The appropriate formula for estimating the least measurable force in the presence of noise is that for the equilibrium state (see section 2.8). Thus if $m = 1$ kg, $T = 300$ K, $\tau^* = 1000$ s, $Q = 60$, $t_0 = 100$ s, $1/\eta = 10$, then from eqn (2.84) the least measurable acceleration limited by thermal noise is 10^{-12} m s^{-2}. This example, not untypical, shows that thermal noise is a fundamental limit to the sensitivity of the beam balance and indicates that the modern beam balance already approaches that limit at room temperature. If the sensitivity of the beam balance is to be improved, the Q factor and fundamental period should both be increased and electrical damping with partial cooling (section 2.10) may be used.

(2) *Temperature effects.* Temperature effects impose severe restrictions on the performance of a beam balance. The change of arm length with temperature will directly produce spurious torques on the balance. It is not possible to make a balance with its two half-arms so similar that the effect of temperature can be exactly cancelled; temperature gradients across the balance arm obviously may make one arm differ from the other (even if they are originally equal). Temperature effects on the beam balance have been considered extensively in the literature of ultraprecision determination of mass but here we wish to understand how temperature affects the measurement of acceleration. Consider a beam balance with two loads (each with mass m) hanging from the ends of the arms of length L. If the length of one arm changes by ΔL it produces a spurious torque equal to $mg\Delta L$, equivalent to an acceleration Δa, given by

$$\Delta a = \frac{\Delta L}{L}g.$$

Thus, for a sensitivity of 10^{-12} m s^{-2}, the relative change $\Delta L/L$ should be less than 10^{-13} during the period of the measurement, a very severe condition. Temperature fluctuations therefore limit the precision of G measurements with the beam balance.

(3) *Tilt of the floor.* Compared to other devices, the effect of tilting of the laboratory floor is particularly severe for a beam balance as any tilt will couple directly to the balance. There are two groups of causes for tilting of the floor, man-made and natural. Even in a quiet place without man-made disturbances, tilting of the floor may reach 10^{-7} rad s^{-2} due to tidal and other geophysical causes (section 3.2).

There are also sources of disturbance to the beam balance, such as magnetic fields, the fluctuations of pressure, the buoyancy effect, and so on, some of which can be reduced by working in a vacuum. Not all disturbances can be so reduced and records of temperature, tilt of the floor, and other quantities, should be made in order to look for correlations between the gravitational results and those environmental factors. It is also

Table 8.2. *Determinations of G with a beam balance*

Authors	Year	Mass (kg)	Form of mass	G (10^{-11} m^3 kg^{-1} s^{-2})
Poynting	1891	153	Sphere	6.67 ± 0.004
Richarz & Krigar-Menzel	1898	10	Block	6.68 ± 0.011
Speake & Gillies	1987	10	Cylinder	6.65 ± 0.23

important to reduce vibration. Not only is it a source of noise, but because the motion of the balance is not strictly linear, vibration may cause parametric deflections of the beam that would appear as the effects of gravitational forces.

Examples of experiments

Table 8.2 gives some results of several laboratory experiments. The experiment of Speake & Gillies (1987), made in the Cavendish Laboratory, was undertaken to examine a new type of beam balance. An 0.80 m long dumbbell beam balance with two attracted masses of 0.499 09 kg and 0.499 47 kg was housed inside a non-magnetic stainless-steel vacuum chamber which was maintained at pressure of 1.3×10^{-5} Pa. A bifilar suspension formed by two parallel suspension wires rather than a knife edge was used to support the beam of the balance; it consisted of two 88 μm diameter, tungsten wires 4.5 cm long separated by 10 mm clamped by two aluminium alloy blocks.

The attracting mass was a 9.187 kg brass cylinder which was attached to the end of a horizontal aluminium alloy rod. The counterweight was also a cylinder, weighing 2.686 kg. The complete assembly was attached at the balance point of the rod to the ceiling through a vertical rod and bearing. A simple mechanism was designed to rotate the system above the attracted masses by pulling a string at a remote place through pulleys. The positional accuracy of the rotating masses was 0.1 mm, which was achieved by rotating the assembly until the horizontal rod gently contacted an aluminium stop which was fixed beforehand. The separations of the attracted masses and the attracting masses were measured using a cathetometer with a precision of ± 0.1 mm.

The calculations of the Newtonian torque were complicated. They included the attractions of the whole assembly of the rotating masses with every part of the dumbbell balance inside the vacuum.

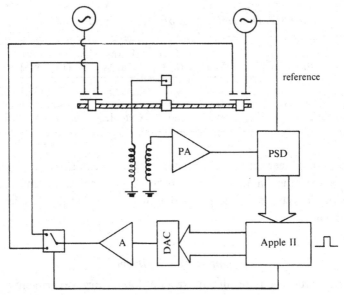

Fig. 8.5 Electrical control circuit in experiment of Speake & Gillies (1987).

Great attention was given to temperature fluctuations. The experimental room was insulated and fitted with a false ceiling and double door; a temperature controller maintained the mean temperature of the room at a value constant to within 0.01 deg per day. The vacuum chamber was surrounded by an outer aluminium alloy case and the intervening volume was filled with polystyrene granules. The daily change in the temperature gradient across the vacuum chamber was 10^{-4} deg, so that the daily change of L due to temperature was within 1 part in 10^9.

An electrical servo-system was used and the gravitational torque on the beam balance was measured by the electric voltage applied to the beam balance to keep the beam as close as possible ($\sim 10^{-9}$ rad) to its equilibrium position. The system is shown in Fig. 8.5. There are two sets of capacitive plates for the beam: detector plates and feedback plates, all semicircular. A.c. voltages ($v \approx 0.3473$ V, $f \approx 17.70$ Hz) of equal amplitude and opposite phase were generated from a biphase oscillator and were applied to the detector plates. The detector plates and other plates in the shape of circles attached to each end of the beam formed two parallel plate capacitors. The deflection of the beam from the null position would cause a current to flow to ground via the beam, the central pivot and the input coil of a transformer as shown in Fig. 8.5. After pre-amplification, the signal was amplified in a phase-sensitive detector (PSD) where the reference signal was provided by the oscillator. The signal from the PSD was read by a computer which calculated the feedback voltage and applied it to the appropriate feedback plate, thus completing the feedback loop. The voltage required to balance the gravitational torque produced by the attracting assembly was read by a Datron, a six and half digits digital voltmeter. The gravitational constant was calculated from the gravitational torque derived from that feedback voltage. In this procedure, careful calibration was necessary to convert the measured voltage to the gravitational torque.

8.3 Direct deflection measurement with torsion balance

Principle

The principle of the use of the deflection method with a torsion balance to measure G is shown in Fig. 8.6. A beam of length $2b$ from which hang two equal attracted masses m is suspended by a thin fibre made of metal or silica. Two attracting masses M are located at distances r from each near attracted mass. Because of the gravitational attraction between M and m, the torsion balance will deflect by an angle θ. The measurement of θ and associated mechanical quantities will give a determination of the gravitational constant.

As an example, if the masses are all spherical and the attracted masses are hung by long enough wires so that the torques on the torsion beam are negligible, then the gravitational constant G can be written as

$$G = \frac{2\pi I r^2}{T^2 bmM}\theta, \tag{8.7}$$

where I and T are the moment of inertia and the period of the torsion pendulum.

There are seven physical quantities to be measured: I, T, b, M, m, r and θ, and all can be measured with a reasonably high precision except the deflection angle θ.

Sensitivity of torsion balance for measurement of G

The sensitivity of the torsion balance has already been discussed in section 4.8. The smallest acceleration is, according to eqn (4.61),

$$\Delta a = \frac{k\Delta\theta}{2bm} = \frac{4\pi^2 b}{T^2}\Delta\theta. \tag{8.8}$$

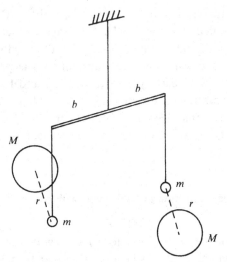

Fig. 8.6 Measurement of G by direct deflection of torsion balance.

For simplicity, the torsion constant k is replaced by $4\pi^2 I/T^2$, and the moment of inertia I by $2mb^2$. It can be seen that, for a given angular resolution, a small torsion balance with a long period of oscillation can give a high sensitivity. The longer the natural period is, the more sensitive the balance will be. In tests of the weak principle of equivalence, the angular resolution should be as high as possible, but there are other considerations for the measurement of G. We see that although the expression for the sensitivity in eqn (8.8) is similar to that in eqn (4.61) for tests of the weak principle of equivalence, the significance is not the same. There are two points worth noting.

(1) It was seen in Chapter 4 that in the experiments testing the weak principle of equivalence, a sensitivity as high as possible should be pursued since that limits the precision of the measurement. However, in measuring G, the overall precision is not determined just by the sensitivity of the torsion balance, but other factors, especially metrological ones, are important. There is no point in having a very high sensitivity without regard to the limits of the metrology. Again, a good match between the sensitivity and the stability of the torsion balance is important. Generally speaking, the sensitivity of the torsion balance is usually sufficient for accurate G measurements; the key issue is not the sensitivity, but the stability, and the higher the resolution, the poorer the stability will usually be. For instance, if it is reasonable to assume that an acceleration of 10^{-8} m s^{-2} may be produced by a

laboratory mass for the torsion pendulum experiment, then for an overall precision of 10^{-4} in G, a sensitivity of 10^{-13} m s^{-2} will suffice. If the angular resolution can reach 10^{-8} rad, and the length b is 0.05 m, a torsion pendulum with a period of 450 s is adequate.

(2) $\Delta\theta$ in the eqn (8.8) is not simply the angular resolution in the sense of eqn (4.61) where $\Delta\theta$ means the minimum detectable angle. In the measurement of G, the gravitational torque is taken to be proportional to θ, and so θ must vary linearly with the torque. In designing the experiment, one must distinguish between an angular resolution which requires linearity in an absolute measurement and an angular resolution which does not have such a requirement in a relative measurement. A detailed discussion follows below.

Design of experiments

High sensitivity is the main merit of the method of the direct deflection, the traditional Cavendish method, but it also suffers from many shortcomings. Great attention should be paid to the following particular points.

(1) *Absolute measurement of angular displacement.* Measuring an absolute angular displacement, particularly when the displacement is as small as 10^{-9}–10^{-10} rad, is difficult. Although modern detectors may have an angular resolution of 10^{-10} rad, or better, absolute measurements need a high degree of linearity in the detector, a matter which is still not fully assured for most types of detector. Suppose, for example, a position-sensitive electro-optic detector gives an output of 1 V for an angular displacement of 10^{-4} rad. If a voltage of 0.1 mV was measured, would the displacement be 10^{-8} rad? Unless the linearity is checked, 0.1 mV would not correspond to 10^{-8} rad. In measurements of G, absolute angle measurements are required and because angles are usually very small, the calibration may be difficult. Clearly it is desirable to have as large a deflection, and therefore as large attracting masses, as possible. Cavendish used attracting masses of 160 kg each in his original experiment. However, at the same time heavy attracting masses give rise to other problems, thus, moving a heavy mass will cause considerable disturbance to the apparatus.

Another way of overcoming the non-linearity of the detector is to use a servo-system to keep the torsion balance close to the equilibrium position by applying a proportional force to the deflection of the balance. Again, careful calibration is necessary and the problem of linearity is moved from the detector to the servocontrol circuits.

(2) *Stability of the equilibrium position.* The stability of a highly sensitive torsion balance has to be considered very seriously. The gravitational torque is balanced by the torque developed by the twisted suspension fibre and, as was seen earlier, that torque is quite sensitive to temperature and also to erratic movements of dislocations in the material of the fibre. The treatment of a fibre to improve the stability of the rest point was discussed in Chapter 3. Changes of local background gravitational attractions are also severe for this sort of experiment. Effects of movements of local objects were also discussed in Chapter 3 for a variety of forms of torsion balance beam but, in experiments for determining G, the torsion balances usually have just a single beam with two masses with just a quadrupole gravitational attraction. Although no use of a framework of higher symmetry has been reported for measurements of G, in principle, one could be used. The calculation of the cross-attraction between the framework and the attracting masses would be especially important for such frames.

(3) *Determination of the separation.* The measurement of the separation between attracted mass and attracting mass is crucial in the determination of G and in practice the measurement is difficult because the attracted ball is usually housed inside a vacuum chamber, and the beam and the attracted mass are freely suspended. It is therefore important to design the experiment so that measurements to the masses on the beam are not necessary but instead, distances between two positions of the large attracting masses are the relevant quantities.

In summary, the direct deflection method using a torsion balance, although very sensitive, has many shortcomings that can be overcome only with elaborate techniques.

Examples of experiments

The first determination of G using the direct deflection of a torsion balance was made by Cavendish (1798). After Cavendish, the most careful work was done by Boys (1895). His work is notable for his use of fused silica as the torsion fibre and for his analysis of the sensitivity of the apparatus which led to a reduction in the size of the balance, very greatly influencing all experimenters after him. His work deserves to be studied by anyone who plans future gravitational experiments.

Before Boys's studies the attracting masses were made as heavy as possible and the beam of the torsion balance as long as possible. Boys (1889) pointed out that this was not the best approach. Higher sensitivity

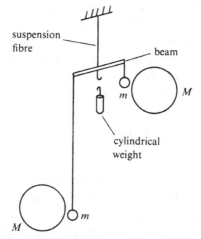

Fig. 8.7 Measurement of Boys (1895).

can be achieved by a torsion balance with a longer natural period, and that depends in part on having a suitable suspension wire. After he had developed fibres of fused silica, he used them to make a torsion balance of very small size to measure the gravitational constant.

Fig. 8.7 shows a diagram of Boys's apparatus. Two balls of gold were hung at the ends of a torsion beam 22.5 mm long. Two sizes of ball were used, one pair was 5 mm in diameter and each ball weighed 1.2983 g; the other pair was 6.3 mm in diameter and each ball weighed 2.6601 g. As the beam was very short, the two balls were hung at different levels, as shown in Fig. 8.7. To reduce the effect of cross-attraction and make the net torque on the beam large enough, the upper ball was 150 mm above the lower ball. The mobile element was suspended by a fine quartz fibre 420 mm long. Two attracting spherical masses of lead weighing 7408.16 g and 7407.47 g respectively were suspended from phosphor–bronze wires. They could be placed in one extreme position shown by the solid lines in Fig. 8.7, or in the opposite position, or they could be removed far away.

Boys paid great attention to the measurements of the seven physical quantities in eqn (8.7) and although some of the techniques he used are now out of date, there are a few points which are worth discussing.

One is the determination of the moment of inertia of the moving system. In principle, the moment of inertia of a rotating assembly may be calculated from the measurements of the dimensions and the masses of the individual parts in the whole assembly but the result will not usually be very accurate because of the imperfection of the geometrical structure and the uncertainty of the rotational axis. To measure the moment of inertia of the moving system, Boys used the method of 'mass replacement'. Initially he measured the period of the balance with the two gold balls fixed at the ends of the beam (period T_1), and then one without the balls at the ends (period T_2). The moment of inertia of the complete assembly (including the suspended masses) is given by

$$I = \tfrac{1}{2}mL^2\left(1 + \frac{T_1^2}{T_1^2 - T_2^2}\right), \qquad (8.9)$$

where m is the mass of the gold ball and L is the length of the beam. However, Boys found that this method was subject to systematic errors, because different tensions in the fibre led to different torsional constants on account of changes in the lengths and diameters of the quartz fibre. Boys then used a cylindrical counterweight of the same mass as the two gold balls and a moment of inertia that could be accurately calculated from its regular shape. The counterweight was hung from a central hook (see Fig. 8.7), stretching to the same extent and preserving the same torsional constant. With this precaution, Boys obtained a value of the moment of inertia to better than one part in 10000. Other methods that give the moment of inertia to high precision will be mentioned below.

Boys paid great attention to the attractions of all masses other than the lead balls. Not only were the forces between all parts of the mobile element and the balls carefully calculated but Boys also made sure that the attractions of the various suspension wires were really negligible. For example, the size of the wire holding the gold balls was carefully chosen so that the attraction between the wire and the lead balls was less than 1/250000 of the attraction between the gold balls and the lead balls.

The procedure that Boys followed was this. When the balance reached its equilibrium position with the attracting masses in one extreme location, the position of the beam was observed and recorded. After the reading, the period of the pendulum was measured, the oscillation of the pendulum being excited by removing and then replacing the attracting masses. The measurement of the period of the oscillation (which was about 193 seconds) was continued until the amplitude of the oscillation was reduced to a very small amount by air damping. The attracting masses were then placed in the opposite extreme location, and the position of the beam was observed and recorded. The above procedure would be repeated as often as necessary to obtain a standard deviation from the mean as small as possible.

Table 8.3. *Boys's (1895) determination of G*

Group	G $(10^{-11} \text{ m}^3 \text{ kg}^{-1} \text{ s}^{-2})$
1	6.6645
2	6.6702
3	6.6711
4	6.6675
5	6.6551
6	6.6579
7	6.6533
8	6.6578
9	6.6695

Table 8.4. *Some results of G measured by the direct deflection method with torsion balance*

Authors	Year		G $(10^{-11} \text{ m}^3 \text{ kg}^{-1} \text{ s}^{-2})$
Cavendish	1798		6.75 ± 0.05
Reich	1852		6.70 ± 0.04
Baily	1843		6.63 ± 0.07
Cornu & Baille	1873		6.64 ± 0.017
Boys	1895	as published	6.658 ± 0.007
		mean of 9 sets	6.664 ± 0.007

It is instructive to see the variation of the measurements, and Table 8.3 gives all Boys's results for the nine trials. Boys's final result was not the average of the above values. He thought that the results of group 6 and group 8 were more consistent and reliable, so he announced his final value as $(6.658 \pm 0.007) \times 10^{-11} \text{ m}^3 \text{ kg}^{-1} \text{ s}^{-2}$. The precision of his experiment is 1 part in 10^3. If all the groups are treated equally, the mean value is 6.664 with a standard deviation of 0.007 or 1 part in 10^3, departing more from the generally accepted CODATA value.

At the conclusion of his work, Boys wrote 'I am still convinced that G can be determined with an accuracy of 1 in 10000 by means of the apparatus such as I have described.'

Table 8.4 gives the results of other experiment for the measurement of G using direct deflection of a torsion balance to compare with that of Boys. If all Boys's sets are given equal weight, it may be seen that the final result

(6.664 ± 0.024) is not significantly different in value or precision from that of Baily and Cornu & Baille.

8.4 Direct deflection measurement with mercury suspension

A test mass of a direct deflection experiment can be supported in other ways than by a beam balance or a torsion balance provided the support exerts a very small restoring force comparable to the gravitation attraction produced by the laboratory mass. De Boer (1987) in the Physikalisch-Bundesanstalt (Germany) floated the test masses on a mercury pool, which because of the absence of friction permits a very sensitive and accurate measurement (see also Keiser & Faller, 1979).

A suspension fibre in the traditional torsion balance experiments has the two functions of giving the balance a fixed equilibrium position and providing a small restoring force, and both those functions have to be provided when the test masses are floated on mercury.

The centering device in the experiment of de Boer *et al.* makes use of the high surface tension of mercury (de Boer *et al.*, 1980). The test masses are fixed to a ring which floats on the mercury and a fixed ring on the surface of the mercury is located close to its inner surface. Any radial shift of the floating ring will extend the mercury surface, and generate a tensional force which will act to centre the ring.

The restoring force is provided by voltages applied to quadrant electrodes with a structure similar to the differential capacitive displacement tranducer shown in Fig. 4.11. Four separate electrodes are used and electrical voltages are applied to them to maintain the equilibrium position. The applied voltage is derived from the deflection of the balance which is measured by a laser interferometer. When the balance is subject to the gravitational force of the attracting masses, the voltage which keeps the balance in its original position is a direct measure of the gravitational attraction.

The attracted bodies were a pair of cylinders made of Zerodur (120 g). The attracting bodies were also a pair of cylinders, but made of tungsten (0.9 kg). The calculation of the gravitational attraction between the cylinders involves considerable computation. The result so far obtained is

$$G = (6.667 \pm 0.005) \times 10^{-11} \text{ m}^3 \text{ kg}^{-1} \text{ s}^{-2}. \tag{8.10}$$

8.5 Torsion pendulum determinations – period method

The high sensitivity of the torsion pendulum means that there is a substantial change of the free period when the pendulum is attracted by large masses, from which the constant of gravitation can be determined.

The method was first used by Reich (1852), then developed by Heyl and co-workers (Heyl, 1930; Heyl & Chrzanowski, 1942) and used to best effect by Luther & Towler (1982).

Principle of period method

The principle of the method is to compare the periods of a torsion pendulum in two different configurations, one with attracting masses near the pendulum (referred to as the 'near' position), and the other with attracting masses away from the pendulum (the 'far' position). Fig. 8.8 shows a typical arrangement of the experiment in the near position with the large masses M_1 and M_2 in line with the equilibrium position of the beam of the pendulum. The attracting masses M_1 and M_2 provide a restoring force additional to that of the torsion fibre, they thus increase the total restoring force on the pendulum masses m_1 and m_2, and hence the frequency of the pendulum. On the other hand, as shown in Fig. 8.9, in the far position, where the masses M_1 and M_2 are at right angles to the equilibrium position of the pendulum, they provide a negative restoring force that reduces the total restoring force and so decreases the frequency of the oscillations. The measurements of the periods in the near and far positions together with the masses and dimensions of the components of the apparatus, enable a value of G to be calculated.

The specific form of the formula will depend upon the detailed configuration of the experiment. The configurations of Figs 8.8 and 8.9 will be used to illustrate the theory of the method. They show a typical experimental set-up and though the formula derived in the following may not apply exactly for different arrangements, the methods of the study are the same. In the near position, shown in Fig. 8.8, the large masses are in line with the pendulum.

Let the moment of inertia of the pendulum with masses M_1 and M_2 at the ends of the beam be I. Let the torsional constant be k. Let V_{ij} be the potential due to the mass M_j ($j = 1$, 2, corresponding to the attracting masses M_1 and M_2 whose shapes are not specified) at the position of mass m_i ($i = 1$, 2, corresponding to the attracted masses m_1 and m_2, which are assumed to be spherical in shape).

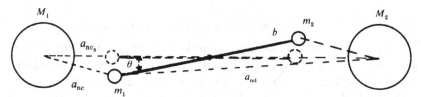

Fig. 8.8 Masses in 'near' position to reduce period of torsion balance.

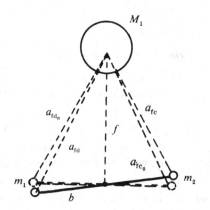

Fig. 8.9 Masses in 'far' position to increase period of torsion balance.

The Lagrangian of the pendulum is then

$$L = \tfrac{1}{2}I\dot{\theta}^2 - \tfrac{1}{2}k\theta^2 - \sum_{j=1}^{2}\sum_{i=1}^{2} m_i V_{ij}. \tag{8.11}$$

In the above formula, $\tfrac{1}{2}k\theta^2$ is the mechanical potential energy of the suspension fibre. In an actual experiment, the potential energy of the beam in the fields of mass M_1 and M_2 must be included but is omitted here to simplify the exposition. Nevertheless, the method of calculating such a contribution will be similar to that now to be discussed. There are four terms in the sum $\sum m_i V_{ij}$, namely: $m_1 V_{11}$, $m_2 V_{22}$, $m_1 V_{12}$, $m_2 V_{21}$, and they may be thought of as direct terms when m_i is close to M_i but cross-terms when m_i is far from M_j. These cross-attractions cannot be neglected even in the near position, and in the far position they are comparable with the direct terms. From the Lagrangian of the system the equation of motion of the pendulum is

$$I\ddot{\theta} + k\theta + \sum_{j=1}^{2}\sum_{i=1}^{2} m_i \frac{\partial V_{ij}}{\partial \theta} = 0. \tag{8.12}$$

We consider first the near position. We assume that the masses M_1 and M_2 are equal, that they are symmetrical with respect to the torsion balance, and that m_1 and m_2 are equal. These assumptions are not necessary, but do represent quite closely typical experiments. The calculation for an asymmetric arrangement will not differ much from this case.

From Fig. 8.8, we may see the following. Let a_{nc} and a_{nd} be the distances of the close ball and distant ball from the centre of one of the attracting masses, let a_{nc_0} and a_{nd_0} be their values when $\theta = 0$, and let b be the length of the half-beam of the torsion pendulum. Then

$$m_1 \frac{\partial V_{11}}{\partial \theta} = m_2 \frac{\partial V_{22}}{\partial \theta} = m_i \frac{\partial V_{ii}}{\partial a_{nc}} \frac{(a_{nc_0} + b)\,b}{a_{nc}} \sin\theta \tag{8.13}$$

and

$$m_1 \frac{\partial V_{12}}{\partial \theta} = m_2 \frac{\partial V_{21}}{\partial \theta} = -m_i \frac{\partial V_{ij}}{\partial a_{nd}} \frac{(a_{nd_0} - b)\,b}{a_{nd}} \sin\theta. \tag{8.14}$$

We use the following approximations:

$$\sin\theta = \theta - \tfrac{1}{6}\theta^3 + \dots, \tag{8.15}$$

$$a_{nc} = a_{nc_0} + \frac{1}{2}\frac{b(a_{nc_0} + b)}{a_{nc_0}}\theta^2 + \dots, \tag{8.16}$$

$$a_{nd} = a_{nd_0} - \frac{1}{2}\frac{b(a_{nd_0} - b)}{a_{nd_0}}\theta^2 + \dots, \tag{8.17}$$

and so

$$\frac{1}{a_{nc}} = \frac{1}{a_{nc_0}} - \frac{1}{2}\frac{b(a_{nc_0} + b)}{a_{nc_0}^3}\theta^2 + \dots, \tag{8.18}$$

$$\frac{1}{a_{nd}} = \frac{1}{a_{nd_0}} + \frac{1}{2}\frac{b(a_{nd_0} - b)}{a_{nd_0}^3}\theta^2 + \dots, \tag{8.19}$$

$$\frac{\partial V_{ii}}{\partial a_{nc}} = \left(\frac{\partial V_{ii}}{\partial a_{nc}}\right)_0 + \left(\frac{\partial^2 V_{ii}}{\partial a_{nc}^2}\right)_0 \frac{b(a_{nc_0} + b)}{2a_{nc_0}}\theta^2 + \dots, \tag{8.20}$$

$$\frac{\partial V_{ij}}{\partial a_{nd}} = \left(\frac{\partial V_{ij}}{\partial a_{nd}}\right)_0 - \left(\frac{\partial^2 V_{ij}}{\partial a_{nd}^2}\right)_0 \frac{b(a_{nd_0} - b)}{2a_{nd_0}}\theta^2 + \dots, \tag{8.21}$$

where $(\partial V_{ii}/\partial a)_0$, $(\partial^2 V_{ii}/\partial a^2)_0$ are the values of the differential $\partial V_{ii}/\partial a$,

$\partial^2 V_{ii}/\partial a^2$ when $\theta = 0$. Then, inserting eqns (8.15) and (8.18)–(8.21) into (8.13) and (8.14), we obtain

$$m_1\frac{\partial V_{11}}{\partial \theta} = m_2\frac{\partial V_{22}}{\partial \theta} = m_i\left(\frac{\partial V_{ii}}{\partial a_{nc}}\right)_0\frac{b(a_{nc_0}+b)}{a_{nc_0}}\theta - \tfrac{1}{2}m_i\left[\left(\frac{\partial V_{ii}}{\partial a_{nc}}\right)_0\frac{b^2(a_{nc_0}+b)^2}{a^3_{nc_0}}\right.$$

$$\left. +\frac{1}{3}\left(\frac{\partial V_{ii}}{\partial a_{nc}}\right)_0\frac{b(a_{nc_0}+b)}{a_{nc_0}}-\left(\frac{\partial^2 V_{ii}}{\partial a^2_{nc}}\right)_0\frac{b^2(a_{nc_0}+b)^2}{a^2_{nc_0}}\right]\theta^3 + \dots, \quad (8.22)$$

and

$$m_1\frac{\partial V_{21}}{\partial \theta} = m_2\frac{\partial V_{12}}{\partial \theta} = -m_i\left(\frac{\partial V_{ij}}{\partial a_{nd}}\right)_0\frac{b(a_{nd_0}-b)}{a_{nd_0}}\theta + \tfrac{1}{2}m_i\left[-\left(\frac{\partial V_{ij}}{\partial a_{nd}}\right)_0\frac{b^2(a_{nd_0}-b)^2}{a^3_{nd_0}}\right.$$

$$\left. +\frac{1}{3}\left(\frac{\partial V_{ij}}{\partial a_{nd}}\right)_0\frac{b(a_{nd_0}-b)}{a_{nd_0}}+\left(\frac{\partial^2 V_{ij}}{\partial a^2_{nd}}\right)_0\frac{b^2(a_{nd_0}-b)^2}{a^2_{nd_0}}\right]\theta^3 + \dots, \quad (8.23)$$

respectively. Retaining just the terms proportional to θ, the equation of motion for small oscillations becomes

$$I\ddot{\theta}+k\theta+k_{ng}\theta = 0, \quad (8.24)$$

where the constant k_{ng} (the gravitational restoring force constant in the near position) is the sum of close and distant terms, namely

$$k_{ng} = 2(k_{nc}+k_{nd}) \quad (8.25)$$

with

$$k_{nc} = m_i\left(\frac{\partial V_{ii}}{\partial a_{nc}}\right)_0\frac{b(a_{nc_0}+b)}{a_{nc_0}} \quad (8.26)$$

and

$$k_{nd} = -m_i\left(\frac{\partial V_{ij}}{\partial a_{nd}}\right)_0\frac{b(a_{nd_0}-b)}{a_{nd_0}}. \quad (8.27)$$

The term k_{nc} is always positive and the close attracting mass in the near configuration provides a positive restoring force; k_{nd} is negative and the distant attracting mass in the near position provides negative restoring force. However, because $a_{nc} \ll a_{nd}$, the absolute value of k_{nd} is much less than that of k_{nc}, so that in the near position the attracting masses provide a net positive restoring force and the period is shorter than when the attracting masses are absent.

The far position is shown in Fig. 8.9 with the attracting masses at right angles to the beam of the pendulum at distance f from the centre of the beam; similarly, we have

$$m_1 \frac{\partial V_{11}}{\partial \theta} = m_2 \frac{\partial V_{22}}{\partial \theta} = m_i \frac{\partial V_{ii}}{\partial a_{\text{fd}}} \frac{fb}{a_{\text{fd}}} \cos \theta, \tag{8.28}$$

$$m_1 \frac{\partial V_{21}}{\partial \theta} = m_2 \frac{\partial V_{12}}{\partial \theta} = -m_i \frac{\partial V_{ij}}{\partial a_{\text{fc}}} \frac{fb}{a_{\text{fc}}} \cos \theta, \tag{8.29}$$

where a_{fc}, a_{fd} are the distances of the close ball and distant ball, respectively, from the centre of one of the attracting masses. As the angle θ is very small, the two distances are almost equal, and the following approximations may be used:

$$\cos \theta = 1 - \tfrac{1}{2}\theta^2 + \dots, \tag{8.30}$$

$$a_{\text{fc}} = a_{\text{fc}_0} - \frac{fb}{a_{\text{fc}_0}}\theta + \dots, \tag{8.31}$$

$$a_{\text{fd}} = a_{\text{fd}_0} + \frac{fb}{a_{\text{fd}_0}}\theta + \dots, \tag{8.32}$$

and so

$$\frac{1}{a_{\text{fd}}} = \frac{1}{a_{\text{fd}_0}} - \frac{fb}{a_{\text{fd}_0}^3}\theta + \dots, \tag{8.33}$$

$$\frac{1}{a_{\text{fc}}} = \frac{1}{a_{\text{fc}_0}} + \frac{fb}{a_{\text{fc}_0}^3}\theta + \dots, \tag{8.34}$$

$$\frac{\partial V_{ii}}{\partial a_{\text{fd}}} = \left(\frac{\partial V_{ii}}{\partial a_{\text{fd}}}\right)_0 + \left(\frac{\partial^2 V_{ii}}{\partial a_{\text{fd}}^2}\right)_0 \frac{fb}{a_{\text{fd}_0}}\theta + \dots, \tag{8.35}$$

$$\frac{\partial V_{ij}}{\partial a_{\text{fc}}} = \left(\frac{\partial V_{ij}}{\partial a_{\text{fc}}}\right)_0 - \left(\frac{\partial^2 V_{ij}}{\partial a_{\text{fc}}^2}\right)_0 \frac{fb}{a_{\text{fc}_0}}\theta + \dots, \tag{8.36}$$

where as before, the suffix 0 denotes values at $\theta = 0$. Therefore, inserting eqns (8.30) and (8.33)–(8.36) into (8.28) and (8.29), we have

$$m_1 \frac{\partial V_{11}}{\partial \theta} = m_2 \frac{\partial V_{22}}{\partial \theta} = m_i \left(\frac{\partial V_{ii}}{\partial a_{\text{fd}}}\right)_0 \frac{fb}{a_{\text{fd}_0}} - m_i \left[\left(\frac{\partial V_{ii}}{\partial a_{\text{fd}}}\right)_0 \frac{f^2 b^2}{a_{\text{fd}_0}^3}\right.$$
$$\left. - \left(\frac{\partial^2 V_{ii}}{\partial a_{\text{fd}}^2}\right)_0 \frac{f^2 b^2}{a_{\text{fd}_0}^2}\right]\theta - \tfrac{1}{2}m_i \left(\frac{\partial V_{ii}}{\partial a_{\text{fd}}}\right)_0 \frac{fb}{a_{\text{fd}_0}}\theta^2 + \dots \tag{8.37}$$

and

$$m_1 \frac{\partial V_{12}}{\partial \theta} = m_2 \frac{\partial V_{21}}{\partial \theta} = -m_i \left(\frac{\partial V_{ij}}{\partial a_{\mathrm{fc}}}\right)_0 \frac{fb}{a_{\mathrm{fc}_0}} - m_i \left[\left(\frac{\partial V_{ij}}{\partial a_{\mathrm{fc}}}\right)_0 \frac{f^2 b^2}{a_{\mathrm{fc}_0}^3}\right.$$

$$\left. - \left(\frac{\partial^2 V_{ij}}{\partial a_{\mathrm{fc}}^2}\right)_0 \frac{f^2 b^2}{a_{\mathrm{fc}_0}^2}\right]\theta + \tfrac{1}{2} m_i \left(\frac{\partial V_{ij}}{\partial a_{\mathrm{fc}}}\right)_0 \frac{fb}{a_{\mathrm{fc}_0}}\theta^2 + \ldots . \quad (8.38)$$

Again retaining just the terms proportional to θ, the equation of motion is

$$I\ddot{\theta} + k\theta + k_{\mathrm{fg}}\theta = 0, \quad (8.39)$$

where the gravitational constant in the far position k_{fg}, is given by

$$k_{\mathrm{fg}} = 2k_{\mathrm{fc}} + 2k_{\mathrm{fd}} \quad (8.40)$$

with

$$k_{\mathrm{fc}} = -m_i \left[\left(\frac{\partial V_{ij}}{\partial a_{\mathrm{fc}}}\right)_0 \frac{f^2 b^2}{a_{\mathrm{fc}_0}^3} - \left(\frac{\partial^2 V_{ij}}{\partial a_{\mathrm{fc}}^2}\right)_0 \frac{f^2 b^2}{a_{\mathrm{fc}_0}^2}\right], \quad (8.41)$$

$$k_{\mathrm{fd}} = -m_i \left[\left(\frac{\partial V_{ii}}{\partial a_{\mathrm{fd}}}\right)_0 \frac{f^2 b^2}{a_{\mathrm{fd}_0}^3} - \left(\frac{\partial^2 V_{ii}}{\partial a_{\mathrm{fd}}^2}\right)_0 \frac{f^2 b^2}{a_{\mathrm{dc}_0}^2}\right]. \quad (8.42)$$

Now in the symmetrical arrangement of Fig. 8.9, $a_{\mathrm{fd}_0} = a_{\mathrm{fc}_0}$, so that the coefficients k_{fc} and k_{fd} are also equal. They are both negative, because for masses of any form, the gravitational attraction is positive, i.e. for an attracting mass with any solid geometrical shape, we will have

$$\left(\frac{\partial V}{\partial a}\right)_0 > 0, \quad (8.43)$$

and it decreases monotonically with distance, i.e.

$$\left(\frac{\partial^2 V}{\partial a^2}\right)_0 < 0. \quad (8.44)$$

Hence, in the far position configuration, the gravitational restoring force is negative and the frequency is decreased.

Summarizing the above study, we can see that the formula for the determination of G may be rewritten as

$$k_{\mathrm{ng}} - k_{\mathrm{fg}} = 4\pi^2 I \left(\frac{1}{T_{\mathrm{n}}^2} - \frac{1}{T_{\mathrm{f}}^2}\right), \quad (8.45)$$

where T_{n} and T_{f} are the periods in the near and far positions configurations respectively.

Sensitivity of torsion pendulum in period method

The smallest acceleration that can be detected by a torsion pendulum in the period method according to the mechanical characteristics of the torsion pendulum is different from that defined in eqn (8.8) of the torsion balance in the direct deflection method. This is because the quantity measured in the period method is time instead of deflection. As for a simple pendulum (see eqn (4.19)), the sensitivity of a torsion pendulum is defined by

$$\Delta a = \frac{8\pi^2}{T^2} b\theta \frac{\Delta T}{T}, \tag{8.46}$$

where ΔT is the time resolution and θ is the swing angle. Evidently the time resolution ΔT of the experiment will determine the resolution of the experiment. However, as we discussed before in section 4.7 for the simple pendulum, it is not the time resolution itself which limits the sensitivity of the torsion pendulum, it is the angular resolution which affects the precision of the 'rest' position of the swing; the effective time resolution is given by

$$\frac{\Delta T}{T} = \frac{\Delta \theta}{n\theta},$$

where n is the number of oscillations in the duration of the measurement and $\Delta\theta$ is the angular resolution for the definition of the 'rest' position. Thus, the resolution of the torsion pendulum in the experiments by the period method is

$$\Delta a = \frac{8\pi^2}{T^2} b \frac{\Delta \theta}{n}. \tag{8.47}$$

It is interesting to compare this sensitivity with that of a simple pendulum (see eqn (4.19)). Roughly speaking, a torsion pendulum is more sensitive than a simple pendulum by a factor of T_t^2/T_s^2, where T_t is the period of the torsion pendulum and T_s is the period of the simple pendulum. The factor could be as high as 10^6 or even higher, and that is why the torsion pendulum is more sensitive than simple pendulum.

To take a practical example, if $b = 0.05$ m, $T = 1000$ s, $n = 10$ and $\Delta\theta = 10^{-10}$ rad, that gives a resolution of 2×10^{-17} m s^{-2}, which is sensitive enough for the measurement of G. For that reason masses in the measurement of G using the period method can be smaller than in the measurement using direct deflection, although, as already pointed out, the precision of a determination of G will usually be limited by other factors.

Design of experiments

As compared with the direct deflection method, the period method has the following advantages.

(1) It is not so sensitive to the drift of the equilibrium position of the pendulum because the quantity measured is time instead of angular displacement. The change of the equilibrium position may affect the measurement of period through the uncertainty of the positions which define the period, but the effect is smaller than on the direct deflection and it is further reduced in proportion to the number of the cycles over which the period is measured. However, the drift of the equilibrium position of the fibre from temperature changes or otherwise is still very important, especially in the far position where the motion may approach instability. It is strongly suggested that the suspension fibre should be properly treated before it is used, as discussed in Chapter 3.

(2) Time measurements are much easier than measurements of angular displacement, not only because of the higher resolution, but also because non-linearity in detectors has far less effect.

(3) Although the torsion pendulum is moving, the quantities that have to be measured are all equilibrium (or stationary) values, in contrast with the direct deflection method, where distances must be measured while the attractions take place. Further if the experiment is designed in a symmetrical way, the final result is rather insensitive to the mean position of the large masses relative to the centre of the beam.

There are at the same time some shortcomings of the method.

(1) *Long measurement time*. The first is the long duration of measurements. Because the period of the torsion pendulum can be, and should be, very long, a single measurement of the period extends over a long time, and so demands good long-term stability of the environmental conditions, such as the temperature of the laboratory, the amplitude of the light source for the optical lever, the electric amplification system, and so on.

(2) *Non-linear effect*. Next there are important non-linear effects in the motion of the pendulum itself, independent of any defect in a detector, but caused by the finite amplitude of the swing and the inhomogeneity of the gravitational field of the large masses (see Chen, 1984a). The non-linear effect is reduced at small amplitudes but the smaller the amplitude, the lower the precision of measurement of the period on account of the uncertainty of the equilibrium position that determines.

As a consequence of non-linear terms in the equation of motion the period is a function of amplitude, not only because the approximation of $\sin \theta$ by θ is not adequate, but also because the torsion pendulum swings in a non-uniform gravitational field produced by the laboratory masses. We can expect that the non-linear effect is much more complicated than that of a simple pendulum where it is in a homogeneous gravitational field produced by the Earth. To show the non-linear effect, consider the same experimental set-up as above. From eqns (8.22) and (8.23), retaining the terms proportional to θ^3 in the equation of motion for the near position, we have

$$I\ddot{\theta} + k\theta + k_{ng}\,\theta + (\alpha_{nc} + \alpha_{nd})\,\theta^3 = 0, \tag{8.48}$$

where

$$\alpha_{nc} = -m_i\left[\left(\frac{\partial V_{ii}}{\partial a_{nc}}\right)_0 \frac{b^2(a_{nc_0}+b)^2}{a_{nc_0}^3}\right.$$
$$\left. + \frac{1}{3}\left(\frac{\partial V_{ii}}{\partial a_{nc}}\right)_0 \frac{b(a_{nc_0}+b)}{a_{nc_0}} - \left(\frac{\partial^2 V_{ii}}{\partial a_{nc}^2}\right)_0 \frac{b^2(a_{nc_0}+b)^2}{a_{nc_0}^2}\right], \tag{8.49}$$

and

$$\alpha_{nd} = m_i\left[-\left(\frac{\partial V_{ij}}{\partial a_{nd}}\right)_0 \frac{b^2(a_{nd_0}-b)^2}{a_{nd_0}^3}\right.$$
$$\left. + \frac{1}{3}\left(\frac{\partial V_{ij}}{\partial a_{nd}}\right)_0 \frac{b(a_{nd_0}-b)}{a_{nd_0}} + \left(\frac{\partial^2 V_{ij}}{\partial a_{nd}^2}\right)_0 \frac{b^2(a_{nd_0}-b)^2}{a_{nd_0}^2}\right]. \tag{8.50}$$

Equation (8.48) is not complete, because damping cannot be ignored since the effect of the non-linear term varies with amplitude. Hence, over a given interval of time the change of period due to non-linearity will depend upon the damping. We therefore study the following equation:

$$I\ddot{\theta} + k\theta + k_{ng}\,\theta + 2\gamma\dot{\theta} + \alpha\theta^3 = 0, \tag{8.51}$$

where γ is the damping factor and $\alpha = \alpha_{nc} + \alpha_{nd}$. It is easy to see that the equation is not satisfied by a damped simple harmonic motion

$$\theta = A\,e^{-(\gamma/I)t}\cos\omega t, \tag{8.52}$$

where A is the amplitude; nor is the period given by the integral

$$T = \int_0^{2\pi} \frac{d\theta}{[(k+k_{ng})\,\theta^2 + \alpha\theta^4]^{\frac{1}{2}}}, \tag{8.53}$$

which applies in the absence of damping. We try instead a solution with a third harmonic term

$$\theta = A\,e^{-(\gamma/I)t}\cos\omega t + B\,e^{-(\gamma/I)t}\cos 3\omega t, \tag{8.54}$$

where B may be expected to be much less than A because the torsion pendulum will mainly oscillate with the fundamental mode even with the presence of non-linear effects.

Inserting eqn (8.54) into (8.51), it is found that

$$\left(-\frac{A\gamma^2}{I}-A\omega^2 I+Ak+Ak_{ng}\right)\cos\omega t+\left(-\frac{B\gamma^2}{I}-9B\omega^2 I+Bk+Bk_{ng}\right)$$

$$\times\cos 3\omega t+\alpha(A\cos\omega t+B\cos 3\omega t)^3 = 0. \quad (8.55)$$

The last term shows that many other harmonic terms beyond the third should be included, in fact an infinite set, but provided B is much less than A, they will be ignored. Because

$$\cos^3\omega t = \tfrac{3}{4}\cos\omega t+\tfrac{1}{4}\cos 3\omega t, \quad (8.56)$$

it follows that eqn (8.55) can be separated into two equations for the coefficients of $\cos\omega t$ and $\cos 3\omega t$ respectively:

$$-\frac{A\gamma^2}{I}-A\omega^2 I+Ak+Ak_{ng}+\tfrac{3}{4}\alpha A^3 = 0, \quad (8.57)$$

$$-\frac{B\gamma^2}{I}-9B\omega^2 I+Bk+Bk_{ng}+\tfrac{1}{4}\alpha A^3 = 0. \quad (8.58)$$

From eqn (8.57),

$$\omega^2 = \frac{k+k_{ng}}{I}-\frac{\gamma^2}{I^2}+\tfrac{3}{4}\alpha A^2. \quad (8.59)$$

Inserting that value into eqn (8.58), the value of B is given:

$$B = \frac{\alpha A^3}{32(k+k_{ng})-32\gamma/I+27\alpha A^2}\approx\frac{\alpha A^3}{32(k+k_{ng})}, \quad (8.60)$$

which is indeed much less than A. Hence the amplitude of the torsion pendulum can be written as

$$\theta_0 = A+B\approx A \quad (8.61)$$

and eqn (8.59) is explicitly

$$\omega^2 = \frac{k+k_{ng}}{I}-\frac{\gamma^2}{I^2}+\tfrac{3}{4}\alpha\theta_0^2+\dots \quad (8.62)$$

or, in terms of the period,

$$T = T_0\left(1+\frac{1}{2}\frac{\gamma^2}{I(k+k_{ng})}-\frac{3}{8}\frac{\alpha}{k+k_{ng}}\theta_0^2+\dots\right), \quad (8.63)$$

where $T_0 = 2\pi I^{\frac{1}{2}}/(k+k_{ng})^{\frac{1}{2}}$ is the period when $\theta_0 = 0$.

Eqn (8.63) is the formula for calculating the period when the non-linear effect is included. There is, to the first order, no direct interaction between the damping and the non-linear effect because the second term in the bracket is constant. But, to the second order, the damping affects the frequency through the decrease in the amplitude.

The non-linear effect is determined by the amplitude and the ratio $\alpha/(k+k_{ng})$. If all parameters are the same except the torsion constant, the larger the torsion constant k is, the less the non-linear effect will be, but at the same time the smaller will be the relative difference of the periods in the near and far positions. Conversely, the greater the difference, and the more effective for the determination of G, the more serious the non-linear effect will be. For a rough estimate, take the potential to be that of a sphere,

$$V = -\frac{GM}{a},$$

with $M = 10$ kg, $a_{nc_0} = 50$ mm, $b = 100$ mm, $m = 50$ g, $T = 1000$ s, and $\theta = 0.4° \approx 10^{-2}$ rad. Since α_{nd} is much less than α_{nc}, the contribution of α_{nd} can be neglected. Then $(k+k_{ng})$ is about 0.4×10^{-7} Nm rad^{-1}, α is about 0.42×10^{-7} Nm rad^{-1}, and the non-linear term is

$$\frac{3}{8}\frac{\alpha}{k+k_{ng}}\theta_0^2 \approx 4 \times 10^{-5}. \tag{8.64}$$

If the aim is to measure G to one part in 10^{-4}, this is appreciable and must be considered in the determination.

The non-linear effect in the far position just cancels out as shown in eqns (8.37) and (8.38), where the terms containing θ^2 have opposite sign and the same value, provided that

$$a_{fc_0} = a_{fd_0},$$

showing that a symmetric design of the experiment can help to reduce the non-linear effect in the far position. If the two arms of the beam are not exactly equal, there will be small non-linear effects but they are unlikely to be significant.

Examples of experiments

There are many famous and successful measurements by the period method to determine G. Table 8.5 lists the results of some of them. Heyl's initial work was published in 1930. The torsion pendulum consisted of two small masses on an aluminium beam (205.9 mm in length, 2.4 mm in diameter) suspended by a tungsten fibre (1 m in length, 2.5 μm in diameter)

Table 8.5. *Determinations of G with torsion pendulum – period method*

Authors	Year	G (10^{-11} m^3 kg^{-1} s^{-2})
Reich	1852	6.59 ± 0.04
Eötvös	1891	6.657 ± 0.013
Braun	1897	6.649 ± 0.002
Heyl	1930	6.670 ± 0.005
Heyl & Chrzanowski	1942	6.673 ± 0.003
Renner	1974	6.670 ± 0.008
Karagioz *et al.*	1976	6.6699 ± 0.002
Sagitov *et al.*	1978	6.6745 ± 0.0008
Luther & Towler	1982	6.6726 ± 0.0005

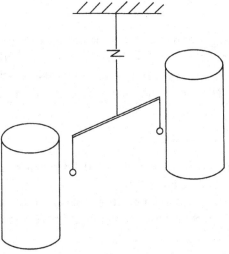

Fig. 8.10 Determination of G by Heyl (1930).

which had been annealed in hydrogen. Three pairs of small masses were used, of gold, platinum and glass, in each case the mass was about 50 g. The assembly of the torsion pendulum was housed in a low vacuum container. The lower portion of the container, where the torsion beam and the small masses were located, was made of brass with several layers of thin sheets of silicon steel for magnetic shielding. To test the effect of the shielding, measurements of the swing periods, were made with an outside magnet alternately magnetized and demagnetized. The large attracting masses were cylinders of steel. Heyl paid attention to possible non-uniformity of the density, and several samples were cut from the material

Table 8.6. *Heyl's (1930) results with different attracted masses*

Materials	G $(10^{-11}\ \mathrm{m^3\ kg^{-1}\ s^{-2}})$	Standard deviation
Gold	6.678	0.003
Platinum	6.664	0.002
Glass	6.674	0.002

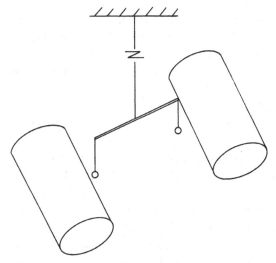

Fig. 8.11 Determination of G by Heyl & Chrzanowski (1942).

from which the large masses were made, to check the variation. He concluded that any effect would be less than 1 part in 50 000. The attracting masses were placed with their axes vertical as shown in Fig. 8.10. The period in the near position was about 1754 s and in the far position, 2081 s. The precision of the measurement of period was about 0.1 s. The results with the three pairs of attracted masses are listed in Table 8.6.

The differences between the materials are outside the range of uncertainties that Heyl quoted, platinum giving apparently a significantly low result. Heyl argued that because the error of the result with gold is one and a half times as great as those with platinum and glass, the three sets of measurement should not have the same weight. The final result quoted in Table 8.5 is obtained by averaging the results of gold, platinum and glass, weighted as 1, 3, 3 respectively.

Heyl & Chrzanowski (1942) performed a subsequent experiment with two main changes. One was the use of photographic recording instead of simple visual observation of the motion of the pendulum. The other was a change in the attitude of the attracting masses: instead of setting the axes of the cylinders vertical, the cylinders were placed with their axes horizontal. In that way the non-linear effect was reduced and the measurement of the distance between the attracting mass and the attracted mass was improved (Fig. 8.11).

Greater attention was also paid to temperature stability. The laboratory, 10 m by 7 m, was divided into two rooms by a partition. The walls of the building were of sheet steel with heat insulation and without windows. Both rooms were heated electrically with thermostatic control. For magnetic shielding, the vacuum case around the pendulum and its case was made of soft iron. Tungsten fibres were again used for the suspension, but two were used, one hard drawn and one annealed. They found that a specially treated tungsten wire may present half the long-term drift of that of an ordinary treated wire, but obtained different values of G with the two types of wire.

Two sets of measurement were made. One was with a fibre of diameter 35.6 μm (0.0014″), the other with one of 30.5 μm (0.0012″) diameter. The periods with the first fibre were 1880 s and 1640 s in the far and near position respectively; with the thinner fibre, 1920 s and 2200 s respectively. The difference corresponds to the fourth power dependence of torque on diameter. The fibre of 35.6 μm diameter was hard drawn and the fibre of 30.5 μm diameter was annealed; with the first, the value was $G = 6.6755 \pm 0.0008$ (10^{-11} m^3 kg^{-1} s^{-2}), with the second, $G = 6.6685 \pm 0.0016$ (10^{-11} m^3 kg^{-1} s^{-2}). The difference, like that between platinum, gold and glass in Heyl's original work, remains without explanation.

The same principle was used by Luther & Towler (1982). Fig. 8.12 is a schematic diagram of the experiment. The torsion pendulum was in a vessel at a pressure 10^{-5} Pa. The suspension fibre was made of quartz and was 10 μm in diameter, 0.4 m in length. It was plated with chromium and gold to make it electrically conducting and to allow the test mass to be grounded. The attracted mass system consists of two tungsten discs mounted in a dumbbell configuration. They were 2.5472 mm thick and 7.1660 mm in diameter on the ends of a centreless-ground tungsten rod 1.0347 mm in diameter and 28.5472 mm long. The mass of the entire assembly was about 7 g. The two attracting masses were tungsten balls, 101.652 72 mm and 101.651 08 mm in diameter with masses 10.489 980 kg and 10.490 250 kg, respectively. The separation of the centres of these two

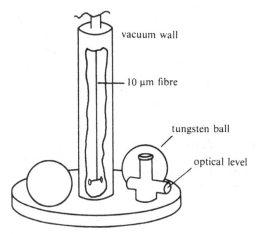

Fig. 8.12 Determination of *G* by Luther & Towler (1982).

attracting masses was 140.059 454 mm when they were in position. Instead of moving the attracting masses to a far position, they were taken away to a great distance. The period without the attracting masses was approximately 360 s and the change in the presence of the two tungsten balls was a small percentage.

The apparatus was in an acoustically isolated, thermally controlled enclosure about 2.5 m³. It in turn was mounted on a concrete slab of about 5 tonnes. The environmental temperature was controlled to within 0.1 deg daily.

The contributions from different sources to the total uncertainty as estimated by Luther and Towler are given Table 8.7, the error of the period and errors of the dimensions of different parts of the attracted mass system are the main ones.

In summary, in the above methods, whether the deflection method or the period method, one can find a common shortcoming: the torsion balance in both methods is subject to external gravitational forces other than the attracting masses. That could be very troublesome as the more sensitive the apparatus is, the longer time the measurement will take, so the greater the external gravitational disturbance will be. As emphasized before, there is no way to shield the unwanted gravitational attraction. However, there are some other methods that are capable of eliminating the effect of extraneous gravitational forces. The reason that the following methods are capable of avoiding a lot of noise is that in all these experiments the true signal and background noise signal are carefully distinguished as 'a.c.' and

Table 8.7. *Contribution to the uncertainty of the measurement of Luther & Towler (1982)*

Source	Uncertainty (parts in 10^6)
Position of attracting masses	10
Mass of attracting mass	1
Length of attracted mass	22
Thickness of attracted mass	36
Density of attracted mass	6
Moment of inertia of mirror	23
Difference of periods	40
Total	64

'd.c.' signals respectively by manipulating the motion of the torsion balance itself or the attracting masses periodically. If the experimental system can be designed with a very narrow bandwidth which is centred around the specific frequency of the true signal, the spurious signal produced by any other source will be located outside the frequency band of the system and will be rejected. Thus, the background noise can be excluded. This is one of the advantages of using laboratory masses, because they can be moved to modulate the desired gravitation signal, making use of the narrow bandwidth to reject the noise.

8.6 Measurement of *G* by angular acceleration

In the direct deflection method *G* is determined by measuring displacements, while the period method determines *G* by measuring velocities. The method discussed in this section, first proposed by Beams *et al.* (1965) determines *G* from accelerations instead of displacements and velocities (see also Beams, 1971).

Principle of angular acceleration method

As shown in Fig. 8.13, one torsion balance is suspended by a thin fibre inside a chamber which is rigidly mounted upon a rotary table with its vertical axis coincident with the axis of the rotation. Two attracting masses outside the chamber are placed on the table as close as possible to the torsion balance. Under the attraction of the attracting masses, the beam of the torsion balance will accelerate towards them. The change of angle is

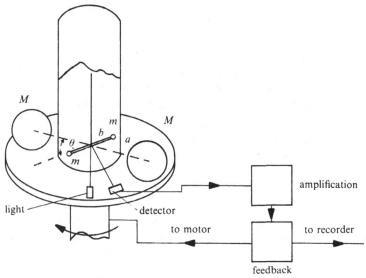

Fig. 8.13 Determination of *G* by angular acceleration.

monitored by an optical lever with the position sensor and the light source all mounted on the table. The signal of the position sensor is then fed back to a servo-motor to drive the rotary table to maintain the angle θ constant. Thus the motion of the balance is transferred to the rotary table. Because the angle θ remains unchanged, the torque acting on the balance is constant and so, therefore, is the angular acceleration of the rotary table. Thus, measuring the change of the angular velocity of the table over a suitable interval of time can give a direct measurement of the gravitational acceleration and, in turn, a precise determination of *G*.

It is easy to see (Fig. 8.13) that if the masses on the table are both spherical, then according to the discussion in section 3.7, the single balance with two arms will be subject to a torque (see eqn (3.38))

$$\tau = 2\frac{GmMb}{a^2}\sin\theta\left[6\frac{b}{a}\cos\theta - 15\frac{b^3}{a^3}(\cos\theta - 7\cos^3\theta) + \dots\right], \quad (8.65)$$

where *M* is the mass of the attracting bodies, *m* is the mass of the attracted bodies (the mass of the beam is ignored here), *b* is the length of the half-arm of the beam and *a* is the distance of one of the attracting bodies from the axis of the rotation. The first term is a quadrupole torque, the second is octupole and so on. If $\theta = \pi/4$, the gravitational quadrupole will reach maximum, so that θ should be set at approximately 45° for the best result.

Sensitivity of angular acceleration method

The method is designed to measure the acceleration through the change of angular velocity of the rotary table driven by the feedback signal from the position sensor. If we assume the gain of the feedback circuit is sufficiently high and the measurement of the change of rate of the rotation is accurate enough, the smallest acceleration which may be detected will be determined by the resolution of the deflection detection system. That means the sensitivity of this method has the same form as that of the direct deflection method of eqn (8.8):

$$\Delta a = \frac{k\Delta\theta}{2bm}, \tag{8.66}$$

where k is the torsion constant of the balance and $\Delta\theta$ is the angular resolution of the detection system. The detector does not have to be linear, only reproducible.

Design of experiments

The novel features of the angular acceleration method are as follows.

(1) The changing rate of the table has to be measured over many revolutions and, consequently, the external gravitational forces are averaged out automatically.

(2) Although the torsion balance is rotating relative to the laboratory, it is stationary with respect to the attracting masses. Hence, the non-linear effect in the period method caused by the motion of the attracting masses relative to the balance, disappears.

(3) The torque acting on the torsion balance is very small, the angular velocity does not change rapidly and hence can be determined quite accurately. It is of course, necessary to continue the observations long enough to obtain good values of differences of velocity but it turns out that the time is not unduly great. For example, according to eqn (8.65), an angular acceleration of 5×10^{-6} rad s^{-2} may be produced by two masses of about 10 kg located within a reasonable distance from the balance. After about two hours, the table will have an angular velocity of one revolution per two minutes. The duration of a measurement is shorter than for the method of periods.

(4) Because it is the acceleration that is measured, spurious constant displacements or constant velocities caused by any disturbance will not affect the final result. In particular any slow creep of the suspension fibre, provided it is constant, will not affect the measurement.

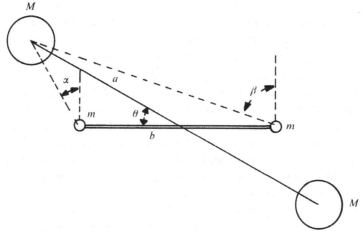

Fig. 8.14 Calculation of the torque on a torsion balance.

Despite these advantages, the angular acceleration method suffers from considerable technical difficulties and only one preliminary determination has been attempted.

The disadvantages are as follows.

(1) *The friction of the bearing.* The friction of the bearing of the rotary table is one of the most troublesome problems, for the residual friction is comparable to the gravitational torque and is not by any means constant. It may vary with the angular velocity, it may also vary with the environmental conditions, temperature, humidity and pressure. Air bearings have low friction but, even with them, the tracking accuracy is no better than ±1°. A magnetic suspension can be almost frictionless, but the problems caused by the magnetic contamination may create further complications. The mercury bearing and associated centering device adopted in the experiment of de Boer (1987, see section 8.4) may provide a better solution.

(2) *High cost.* The apparatus is quite complicated mainly because it incorporates a very precise rotary table which needs not only to be frictionless and non-magnetic, but should also be well isolated from vibration. Such a rotary table is very costly to design and make.

(3) *Optimum position for the attracting masses and the attracted masses.* The attracting masses are quite close to the balance and the ratio of b to a is a number which is indeed less than 1 but sufficiently close to 1 that the

polynomial terms in eqn (8.65) do not converge rapidly. The higher order gravitational multipole terms are not negligible. In fact, a closed formula should be used to calculate the torque so that the optimum angle θ_0 for the maximum torque can be derived. This angle (not necessarily 45°) will be a function of the relative position of the masses. The optimum θ_0 may help in finding the right location of the attracting masses in the experiment as was explained in the use of the stationary point of a hollow cylinder.

We give a simple example of calculating the torque for a simplified system, shown in Fig. 8.14, where attracting masses M and attracted mass m are both assumed to be of spherical shape and the mass of the beam is ignored. The torque exerted on the balance by the masses is given by

$$\tau = \frac{2GMb\cos\alpha}{a^2 + b^2 - 2ab\cos\theta} - \frac{2GMb\cos\beta}{a^2 + b^2 + 2ab\cos\theta}, \tag{8.67}$$

where (see Fig. 8.14)

$$\cos\alpha = \frac{a\sin\theta}{(a^2 + b^2 - 2ab\cos\theta)^{\frac{1}{2}}}, \quad \cos\beta = \frac{a\sin\theta}{(a^2 + b^2 + 2ab\cos\theta)^{\frac{1}{2}}}. \tag{8.68}$$

Thus we have

$$\tau = 2GMab\sin\theta \left[\frac{1}{(a^2 + b^2 - 2ab\cos\theta)^{\frac{3}{2}}} - \frac{1}{(a^2 + b^2 + 2ab\cos\theta)^{\frac{3}{2}}} \right]. \tag{8.69}$$

Then the angle θ_0 which gives the maximum torque is determined by the following equation:

$$\frac{\cos\theta_0(a^2 + b^2) - 3ab + ab\cos^2\theta_0}{(a^2 + b^2 - 2ab\cos\theta_0)^{\frac{5}{2}}} - \frac{\cos\theta_0(a^2 + b^2) + 3ab - ab\cos^2\theta_0}{(a^2 + b^2 + 2ab\cos\theta_0)^{\frac{5}{2}}} = 0. \tag{8.70}$$

The value of θ_0 can easily be found numerically. Fig. 8.15 shows the relation between the optimum angle θ_0 and ratio a/b and we can see that only when the ratio a/b is large is θ_0 equal to 45°.

Examples of experiments

Rose *et al.* (1969) have performed the only experiment. A small accurately made cylindrical rod was the attracted mass. It has a moment of inertia of 4.1×10^{-7} kg m² and was suspended inside a metal chamber by a quartz fibre of 25 μm diameter and 0.33 m long. The attracting masses were two

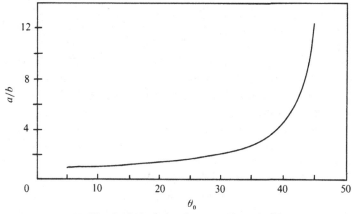

Fig. 8.15 Relation between θ_0 and a/b.

tungsten spheres 101.6 mm diameter, having masses of 10.48998 ± 0.00007 kg and 10.49025 ± 0.00007 kg. Departures from sphericity were less than 12×10^{-6} cm. The attracting masses used in the experiment of Rose *et al.* were those used in the experiment of Luther and Towler. Each rested on a three-point support which in turn was supported by a common quartz plate. The metal chamber was filled with helium at atmospheric pressure. The rotary table was mounted on a cement slab in a clean room with temperature controlled to within 0.5 deg per day. With the optical lever and servo-system as described before, the tracking error was less than ±1 arcsec.

The time required for each revolution of the table was determined by a quartz clock which was reliable to better than 1 μs. In most of the experimental runs, the rotary table was started from rest with the attracting masses in place and $\theta = 45°$. The acceleration was between 4×10^{-6} and 5×10^{-6} rad s^{-2} and it required about 30 min for the first revolution and 2 hours to reach the angular speed of 0.5 rpm.

After the desired speed was reached and the acceleration was measured, the attracting masses were removed to measure the residual acceleration caused by any of the other attractions of the masses on the rotary table. This was usually about 0.1×10^{-6} rad s^{-2}. After subtracting this residual acceleration, the pure acceleration due to the attracting masses could be determined.

The experiment was also performed in the opposite way by giving the table an initial velocity and arranging that the gravitational acceleration reduced the angular velocity. The preliminary result of the experiment was $G = (6.674 \pm 0.012) \times 10^{-11}$ N m^2 km^{-2}.

As we know, the gravitational attraction is extremely weak on the laboratory scale, and thus the amplitude of the displacement of the deflection of the torsion balance is very small. The smallness of the amplitude is usually the reason for measurement errors. As the fourth type of experiment to determine G, we consider a different experimental approach which can greatly enlarge the displacement of the torsion balance and raise the ratio of signal to noise by resonance.

8.7 Resonance method

Principle

Because gravitational forces are weak and static displacements of detectors, such as the torsion balance, are small, there should be considerable advantage in the use of a resonant system to increase the amplitude of displacement and hence the signal-to-noise ratio. The use of mechanical resonance to produce a large amplitude of oscillation by a weak periodic force was first suggested by Joly (1890) who proposed that G might be determined by observing the oscillation of a simple pendulum driven by another identical simple pendulum coupled to it by gravitational attraction. His idea was not put into practice because of the insensitivity of the simple pendulum but has been realized with torsion pendulums.

Consider a torsion pendulum which is swinging under a periodic gravitational torque. Its equation of motion will be

$$I\ddot{\theta} + 2\gamma\dot{\theta} + k\theta = \tau_0 e^{-i\omega t}, \tag{8.71}$$

where I is the moment of inertia, γ is the damping factor, k is the torsion constant and τ_0 is the amplitude of the gravitational torque. When the frequency of the forcing acceleration ω and the natural frequency of the torsion pendulum ω_0 have the following relation

$$\omega^2 = \omega_0^2 - 2\frac{\gamma^2}{I^2}, \tag{8.72}$$

the motion will be resonant with an amplitude

$$A = \frac{\tau_0/I}{2\gamma(\omega_0^2 - \gamma^2/I^2)^{\frac{1}{2}}}. \tag{8.73}$$

The amplitude is determined by the ratio γ/I and will be much greater than

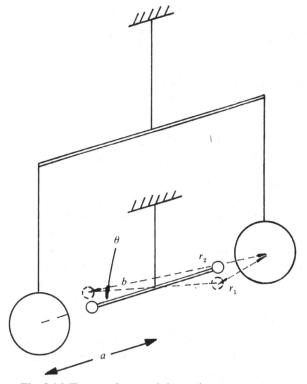

Fig. 8.16 Two torsion pendulums close to resonance.

τ_0/k, the value in the static state, if γ is small enough. To see this explicitly, we write

$$\omega_0^2 = \frac{k}{I},$$

$$\gamma = \frac{2I\omega_0}{Q},$$

and

$$\tau_0 = k\theta_0,$$

where Q is the quality factor of the pendulum and θ_0 is the deflection angle in a static experiment. Because $\omega_0^2 \gg \gamma^2/I^2$, we have the following simple result for A:

$$A = \frac{Q\theta_0}{4}. \tag{8.74}$$

This gives a clear picture of the efficiency of the resonant method, that is, the amplitude of a torsion pendulum in a resonant state will be about $Q/4$ times the amplitude in a static state for the same magnitude of the external torque. $Q = 10^3$ is a very reasonable estimate for a small torsion pendulum in vacuum, which means that it should be possible with a resonant method to have the masses of the attracting bodies some 250 times less than those in a static experiment of the same sensitivity.

Of course, the problem in experiments is how to make a periodic gravitational torque couple to the torsion pendulum in a convenient and efficient way. In the following, three configurations will be discussed.

Three configurations of coupling

Double torsion pendulum

One of the coupling methods uses two torsion pendulums, as shown in Fig. 8.16. The primary one has two attracting masses M close to two attracted masses m attached to a smaller torsion pendulum, and is an active pendulum. The four masses are set at the same horizontal level. If the active torsion pendulum is set swinging with an angle θ, as shown in Fig. 8.16, the torque acting on the passive torsion pendulum (which initially starts from rest) is

$$\tau = 2GMmba \sin \omega t \left(\frac{1}{r_1^3} - \frac{1}{r_2^3} \right),\tag{8.75}$$

where r_1 and r_2 are the distances of close and distant attracted masses to the mass M, as shown in Fig. 8.16. The torque, τ, is not a strictly sinusoidal function of time because the value of $(r_1^{-3} - r_2^{-3})$ is not constant with respect to time. The coupling torque should be as close to a sinusoid as possible, because only the first harmonic component of torque drives the passive pendulum at resonance. If the beams of the torsion pendulums are long enough, and the amplitude of the oscillation of the active pendulum is small enough (assumptions that are reasonable in real experiment conditions), then

$$\frac{1}{r_1^3} \gg \frac{1}{r_2^3} \quad \text{and} \quad \theta \ll 1,$$

so that eqn (8.75) takes the approximate form

$$\tau = \frac{2GMmab}{(a-b)^3} \sin \omega t \left(1 - \frac{3}{2}\frac{b^2}{a^2} + 3\frac{b}{a}\cos \omega t + \dots \right).\tag{8.76}$$

The ratio of the second harmonic torque to the first harmonic torque is

$$D = 3\frac{b}{a},\tag{8.77}$$

so provided that b/a is much less than 1, the coupling torque can be regarded as sinusoidal.

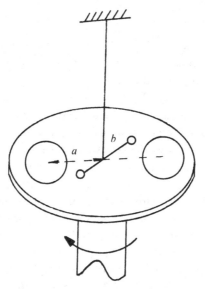

Fig. 8.17 Pair of masses rotating around a torsion pendulum in resonance.

Attracting masses rotating around the torsion pendulum

Figure 8.17 shows another form of coupling: two symmetrically placed attracting masses are located on a round table rotating uniformly around a torsion pendulum. The rotational axis of the table is designed to be coincident with the suspension wire of the torsion pendulum. As before, for a high coupling efficiency the four masses are located in the same horizontal plane. Evidently, the frequency of the rotary table should be half that of the torsional pendulum at resonance. It is found that the torque acting on the torsion pendulum with spherical rotating attracting masses M is

$$\tau = 2GMab\sin\Omega t\left[\frac{1}{(a^2+b^2-2ab\cos\Omega t)^{\frac{3}{2}}}-\frac{1}{(a^2+b^2+2ab\cos\Omega t)^{\frac{3}{2}}}\right],$$

$$(8.78)$$

where a is the distance between M and the rotational axis, b is the half-length of the beam of the torsion pendulum and Ω is the angular velocity of the rotary table. Expanding τ in powers of b/a (less than 1), we have

$$\tau = \frac{GMmb^2}{a^3}\left[6\sin 2\Omega t-105\frac{b^2}{a^2}(1+\cos 2\Omega t)-30\frac{b^2}{a^2}\cos^2\Omega t+\dots\right]. \quad (8.79)$$

Evidently, b^2/a^2 must be much less than 1 if the torque due to the rotating attracting masses is to be sinusoidal in time.

Summarizing the two methods above, it seems that one problem is that the forced gravitational torque does not vary sinusoidally in time. To obtain a good approximation to a sinusoidal torque, the ratio b/a should be very small, but that is difficult because in practice the attracted mass must be as close as possible to the attracting mass to produce a large torque. The condition for a sinusoidal form of torque in the second method is that the ratio $(b/a)^2$ should be very small; a less stringent condition than the former, but still unrealistic, both because it is rather inelegant to use a large rotary table and because the gravitational attraction will be reduced as the distance a increases. However, the higher harmonics do not seem to be very important. If the equation of motion of the driven pendulum is linear, any force at a higher harmonic can be ignored, though that may not be so if the motion is non-linear. Another common disadvantage of both methods is the weakness of coupling. The amplitude of the torques are proportional to $1/(a-b)^3$ and $1/a^3$ respectively in the first and second methods instead of being $1/(a-b)^2$ in the ordinary experiment. Thus, a pair of relatively large attracting masses must be used in the experiments.

There is, however, a third way of coupling the attracting mass to the driven pendulum which may be more satisfactory.

An attracting mass rotating around the attracted mass

As shown in Fig. 8.18, a torsion pendulum has one mass hung far below the beam while the other, the counterweight, is on the beam. A single attracting mass M is located on a rotary table which rotates around the attracted mass m hanging from one end of the beam. Because the counterweight is fixed on the beam at a different level, the unwanted cross-attraction between it and the attracting mass should be negligible. When the attracting mass rotates with a uniform angular velocity ω, the torsion pendulum will be subject to a sinusoidal forcing torque given by

$$\tau = \frac{GMmb}{a^2}\sin\omega t, \qquad (8.80)$$

if M and m are spheres and a is the distance between them.

There are two problems in using the above scheme in practice. One is the cross-attraction of the counterweight and the beam. In fact the cross-attraction has both fundamental and higher harmonic terms and the length of the wire l that supports the mass m should be chosen so that the higher harmonic components of the cross-attraction are sufficiently small.

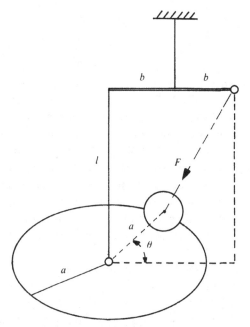

Fig. 8.18 Single mass rotating around a torsion pendulum in resonance.

Using the symbols in Fig. 8.18, the gravitational torque on the beam from the cross-attraction is

$$\tau' = -\frac{GMmab \sin \theta}{(l^2 + a^2 + 4b^2 - 4ab \cos \theta)^{\frac{3}{2}}}$$

$$= \frac{GMmab \sin \theta}{l^3}\left(1 - \frac{3a^2}{2l^2} - \frac{8b^2}{l^2} + \frac{6ab}{l^2}\cos \theta + \dots\right). \quad (8.81)$$

From eqn (8.80), the ratio of the cross-torque τ' to the direct torque τ is

$$\frac{\tau'}{\tau} = \frac{a^3}{l^3}\left(1 - \frac{3a^2}{2l^2} - \frac{8b^2}{l^2} + \frac{6ab}{l^2}\cos \theta + \dots\right). \quad (8.82)$$

The first three (constant) terms in the bracket give the ratio of the fundamental components, so that the ratio of the second harmonic torque to the first harmonic torque is

$$D \approx \frac{6a^4 b}{l^5}. \quad (8.83)$$

It is much smaller than those for the other two methods because it involves the length of the suspension l, and provided l is greater than a and b, D will be very small. In this scheme it is particularly important to suppress the higher-order harmonic components because they may excite the fundamental mode of the oscillation through the non-linear character of the torsion pendulum and cause extra motion. Suppose it is decided that the non-sinusoidal torque produced by the counterweight should be less than 5×10^{-5}, and that b is chosen to be 1 cm and a to be 4 cm; then $l = 30$ cm will be sufficient. This example shows that a pure simple harmonic gravitational torque can be obtained with a reasonable design. The design does not, however, remove the cross-terms in the sinusoidal torque, which are represented by the constant factors in $|\tau'/\tau|$ (see eqn (8.82)) and are of order a^2/l^2. With the same dimensions as in the above example, the cross-torque between the attracting mass and the counterweight is about 1/500 of the torque between the attracting mass and the attracted mass calculated by eqn (8.80).

Vibration caused by the uneven load on the rotary table may also be a problem. Because there is only one attracting mass, the table will tilt as it rotates and so may excite the pendulum directly because the frequency of the tilt is the same as the resonant frequency. The problem may possibly not be too serious because the gravitational efficiency is high, and the attracting mass can be quite small, perhaps 0.1–1.0 kg. If the rotary table is made heavy and rigid, a mass as small as that would not cause much vibration, particularly if there is a good isolation between the torsion pendulum and the rotary table. However, these remarks show, as has been mentioned elsewhere, that the apparent advantages of some design of experiment (here, the high sensitivity) may have to be severely curtailed to avoid other difficulties (tilting of the rotary table). It is for such reasons that the true uncertainties of determinations of G all seem to be of same order, whatever the method.

Design of the experiments

The merits of the resonance method are as follows.

(1) *High gravitational efficiency.* Generally speaking, a torsion pendulum has two important features: high sensitivity and a high mechanical quality factor Q. Most methods only take advantage of the first. The resonance method, however, takes advantages of both; in particular, it makes use of the high Q of a torsion pendulum to increase the gravitational efficiency and reduce the bandwidth which is also associated with the value of Q.

The gravitational efficiency in the resonance method depends on the means of coupling. The double torsion pendulum gives the worst efficiency because the swing angle is very small and the torque is inversely proportional to r^3. The second method is restricted because the attracting mass cannot be close to the attracted mass and because there is very strong cross-attraction with the other attracted mass at the opposite end of the torsion beam. The third method may give the highest efficiency because the attracting mass can be located as close as in the method of direct deflection, the torque is inversely proportional to r^2 and also the cross-attraction (which only has a component approximately proportional to a^3/l^3, as shown in eqn (8.82)) may be reduced to a low level. An attracting mass as small as 100 g might suffice for a measurement of G.

In the history of laboratory measurement of G, the mass of attracting bodies has ranged from 10^5 kg (Richarz & Krigar-Menzel, 1898), to 160 kg (Cavendish, 1798), to 10 kg (Heyl, 1930; Heyl & Chrzanowski, 1942), to 0.9 kg (de Boer, 1987) and to possibly 100 g, an interesting picture of the spectrum of the mass used in this oldest branch of physics. The wide range of this spectrum and the close values of G obtained from them also gives convincing evidence for the validity of the inverse square law in the laboratory scale, a topic which has been discussed in Chapter 7.

If a small mass could be used, not only would vibration be reduced but it would be easier to manufacture, and it would be possible to use a body of low density. That would be important. A possible inhomogeneous distribution of density of the attracting mass is one of the troublesome problems in the determination of G. If a transparent material, such as optical glass or fused quartz, with a very uniform density, could be used and the distribution checked optically, that problem would be much alleviated. Further, fused silica can be worked to a very precise geometrical form and has a much smaller temperature coefficient than most metals.

(2) *Narrow bandwidth.* A torsion pendulum in a resonance method may have a very narrow bandwidth because the Q of the system can be high. That will limit noise, if the frequency of the noise is outside this bandwidth.

For a second-order differential equation such as eqn (8.71) the power response of the amplitude in respect of the frequency ω is

$$P(\omega) = \frac{\tau_0^2/I^2}{(\omega^2 - \omega_0^2)^2 + 4\gamma^2\omega^2/I^2}, \tag{8.84}$$

where τ_0, I, ω_0 and γ are defined in eqns (8.71) and (8.72). This power spectrum is the same as that of eqn (2.79). Therefore, following the same arguments from eqns (2.79)–(2.83), the bandwidth should be about

$$\Delta f = f_0\left[\left(1+\frac{2}{Q\eta^2}\right)^{\frac{1}{2}}-\left(1-\frac{2}{Q\eta^2}\right)^{\frac{1}{2}}\right] \approx \frac{2f_0}{Q\eta^2}, \qquad (8.85)$$

where the meaning of the dimensionless coefficient η^2 is such that $\eta^2 A$ is the variation of the peaks of the motion of the pendulum due to noise. If, for example, $Q = 10^4$ and $\eta^2 = 10^{-3}$, then Δf is about $0.2f_0$. With such a bandwidth torques of higher harmonic order ($2f_0, 3f_0, \ldots$) and most of the external noise will be rejected. The narrow bandwidth is also effective in eliminating the gravitational effects of local moving bodies and provides a 'gravitational filter'.

(3) *Short measurement time.* As the gravitational efficiency is high, the period of the torsional pendulum does not have to be very long. If the third coupling method is employed, a torsion pendulum with period of 1–2 min will be adequate and that would make the whole measurement procedure shorter and more efficient.

We now discuss some particular points in the design. The discussion will specifically consider the third method, but for the most part the conclusions apply to all coupling methods in resonance experiments.

(1) *Interference between the transient oscillation and the resonance oscillation.* The solution of the equation of motion of a torsional pendulum contains a transient as well as a forced term:

$$\theta = a\,e^{-\gamma t/I} \cos \omega_0 t + A \sin (\omega t + \phi), \qquad (8.86)$$

where ϕ is the phase difference between the forced torque and the motion of the pendulum. The first term is the transient oscillation, which decays with the time constant I/γ and will go to zero after an elapsed time. Now the Q of the pendulum is made as high as possible to optimize the performance, but a very high Q will entail a long time constant of the transient motion. If the amplitude of the transient oscillation is not zero, this residual oscillation will couple with the steady-state oscillation and produce an interference beat at a frequency equal to the difference of ω and ω_0, which may lead to difficulties in the measurement.

Consequently, the pendulum should be damped before the forcing torque is applied to the pendulum to start the experiment, and after the transient oscillation has died out the damping may be removed when θ and $\dot{\theta}$ are both zero.

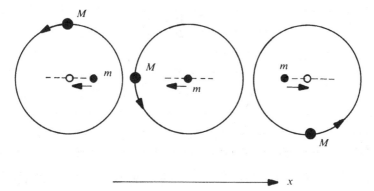

Fig. 8.19 Relative positions of masses M and m; mass m attached to the end of the torsion pendulum can move only in the x-direction, mass M moves in a circular orbit.

However, even if the pendulum is initially at rest, the natural frequency may be excited if the pendulum is not properly isolated. Vibration isolation, particularly for those vibrations with frequency close to the natural and resonance frequencies, is essential.

(2) *Motion of the attracted mass.* So far we have considered a rather ideal oscillator for which the amplitude of the forcing torque τ_0 (eqn (8.71)) is constant. However, this is not strictly true, because at resonance the attracted body will move so that the instantaneous distance between the attracting mass and the attracted mass will then vary with time and in turn make the amplitude of the forcing torque vary with time. This variation will of course make the torque irregular. The effect will depend on coupling methods and in particular on how the pendulum and masses are suspended. We take the third method as an example. We first observe that off exact resonance there is a phase difference ϕ between the forcing torque and the motion of the beam given by

$$\tan \phi = -\frac{2\gamma\omega}{(\omega_0^2 - \omega^2) I}. \tag{8.87}$$

The negative sign means that the motion of the beam lags behind the driving torque. At exact resonance the phase difference is given by

$$\tan \phi = -\frac{Q\omega}{2\omega_0}. \tag{8.88}$$

When Q is large, ϕ is about $-\tfrac{1}{2}\pi$, so that when the attracting mass M is in

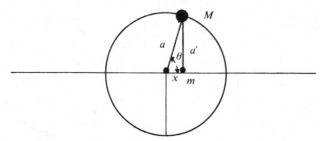

Fig. 8.20 Instantaneous distance between attracting mass M and attracted mass m.

the position where its attraction is maximum, m is in its equilibrium position, whereas if M is in the position where the torque is zero, m is displaced to its maximum (see Fig. 8.19). Evidently, the phase difference will affect the attraction of the moving mass. We now study this problem analytically.

Fig. 8.20 shows the instantaneous position of M and m, and the instantaneous distance between M and m is given by

$$a'^2 = a^2 + A^2 \sin^2(\theta + \phi) - 2aA \sin(\theta + \phi) \cos \theta, \qquad (8.89)$$

where

$$\theta = \omega t.$$

Thus, instead of eqn (8.80), the instantaneous torque exerted by the attracting mass M on the pendulum is

$$\tau' = \frac{GMmb}{a^2 \left[1 - 2\dfrac{A}{a}\sin(\theta + \phi)\cos\theta + \dfrac{A^2}{a^2}\sin(\theta + \phi) \right]} \sin\theta. \qquad (8.90)$$

Because of the condition

$$\frac{A^2}{a^2} \ll \frac{A}{a} \ll 1, \qquad (8.91)$$

eqn (8.90) can be rewritten as

$$\tau' = \frac{GMmb}{a^2}\left[1 + 2\frac{A}{a}\sin(\theta + \phi)\cos\theta + \dots \right]\sin\theta \qquad (8.92)$$

or, expanding the trigonometrical terms, and putting $\theta = \omega t$,

$$\tau' = \frac{GMmb}{a^2}\left[\sin\omega t + 2\frac{A}{a}\cos(\omega t - \phi) - \frac{1}{2}\frac{A}{a}\cos(3\omega t + \phi) + \dots \right]. \qquad (8.93)$$

The third term in the bracket is the higher frequency component, but with the narrow bandwidth of the resonant pendulum, that will not have any serious effect. The second term has the same frequency as the resonance frequency and needs some attention. First, we notice that it depends on the amplitude, which means that in calculating the torque, it must be included if the amplitude is not small. It differs in phase from the principal term, but the phase difference is usually very small, indeed if Q is very large, ϕ will be $-\frac{1}{2}\pi$, so that $\cos(\omega t - \phi)$ will be $\sin \omega t$, so making the calculation straightforward.

(3) *Magnetic contamination.* Should there be any moving electrical parts or electrical currents at the resonance frequency, there may be serious magnetic disturbances of the torsion pendulum. Therefore, any magnetic influence on a resonant experiment will become particularly troublesome if an electric motor is used to drive the rotating table used in the second or third method. Of course, there may be other ways to drive the table without using an electric motor (a hydraulic power drive, for example), but an electric motor is the most convenient. The magnetic influence may come not only directly from the driving motor, but also from the bearings which are usually made of steel. The non-uniform distribution of the residual magnetization of the main bearing carrying the table will couple directly to the torsion pendulum. This coupling could be very strong, because the variation of the magnetic field of the motor and the bearings has the same frequency as the resonant frequency. Therefore, as an important precaution, the driving motor must be located far away from the torsion pendulum and the bearings used in the rotary table should be made of non-magnetic materials – plastic bearings are particularly recommended.

Examples of experiments

Zahradnicek (1933) used a double pendulum method to measure G in a resonant experiment. The primary pendulum consisted of a steel wire supporting a beam with two heavy spheres at each end (see Fig. 8.16). The second beam was 40 cm long with two small lead spheres at the ends and was suspended by a platinum–iridium wire. When the primary pendulum started to swing, the attracting masses produced a gravitational torque on the internal one which built up a considerable amplitude.

Facy & Pontikis (1970, 1971) used the second coupling method rotating two spheres around a torsion pendulum. Their result was $G = (6.6714 \pm 0.0006) \times 10^{-11} \text{ m}^3 \text{ kg}^{-1} \text{ s}^{-2}$. Their experimental set-up is shown in Fig. 8.21.

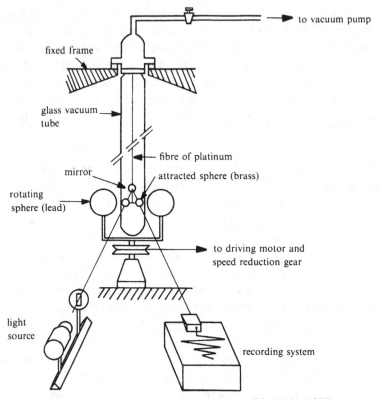

Fig. 8.21 Experiment of Facy & Pontikis (1970, 1971).

Chen has done some initial work with the third coupling method (see Chen, 1984b) which is being continued at the Huazhong University of Science and Technology.

8.8 Conclusion

The results of determinations of the gravitational constant undertaken within the last two decades are given in Table 8.1, and it can be seen that the overall precision is quite modest, about 4 parts in 10^3 for any of the results. The overall mean is 6.6705×10^{-11} N m^2 kg^{-2} with a standard deviation of 0.009×10^{-11} N m^2 kg^{-2}, compared with the value adopted in the CODATA system of 6.6725×10^{-11} N m^2 kg^{-2} with a standard deviation of 0.00085×10^{-11} N m^2 kg^{-2}.

The direct measurement of the deflection of a torsion balance or other detector is clearly inferior to the measurement of the change of period of an oscillator, but a study of the many determinations suggests that one of the main difficulties in any of them is the calculation of the attractions between all parts of the apparatus and not just those which constitute the main masses. Boys indeed demonstrated the advantages of a small apparatus, but the smaller the apparatus the more significant are cross-attractions and the attractions of the framework that supports the mobile masses. The root of the matter is probably not the actual numerical work, which can now be done by computer, but the measurement of the various auxiliary components, the determination of their masses, and the definition of their positions when oscillating. It seems clear that if the measurement of the constant of gravitation is to attain a precision anywhere near that of the elementary measurements of mass, length and time, then the apparatus has to be designed so far as possible to improve the definition of the auxiliary attractions, as well as using masses of forms for which the attractions are as simple to calculate as possible and approach most closely to those of spheres.

9

Conclusion

We have not dealt in this book with all possible experiments on gravitation that have been or could be carried out in the laboratory, whether on the ground or in a space vehicle, but have concentrated on those on which most work has been done and from which most results have been obtained. That is because we have been concerned more with questions of experimental design and technique rather than with the bearing of the results on theories of gravitation. Something was said of that in the Introduction and we simply call attention again to recent reviews such as those of Cook (1987b), Will (1987) and others in the Newton Tercentary review of Hawking & Israel (1987). We have restricted our accounts to the weak principle of equivalence, the inverse square law and the measurement of the constant of gravitation partly because in numbers of results they dominate the subject, but more importantly because, having been so frequently and thoroughly studied, it seems that all the significant issues of experimental method and design are brought out when they are considered.

It was observed in the conclusion of the last chapter on the constant of gravitation, that the definition and calculation of the entire attraction upon a detector such as a torsion balance is no simple matter, and that applies equally to experiments on the inverse square law, as may be shown by the details of the calculations that were necessary in the experiments of Chen *et al.* (1984). Again, if the attractions are to be estimated reliably, it must be established that the density of any body is uniform, and that may be no straightforward matter. It is also necessary to consider carefully the properties that the various masses should have. In some instances it is desirable that the attraction should be insensitive to position in order to minimize some difficult problem of metrology, while in others it may be best to choose a form of body of which the attraction is as close as possible to that of a sphere, having minimal components of higher harmonic order. The investigations described in Chapter 6 enable bodies to be designed to meet such desiderata.

The measurement of distance *per se* to a precision of one part in a million or so is nowadays easily achieved; the problem in experiments on gravitation is to define physically the distances to be measured and to ensure that they correspond to the parameters of the model used to calculate the results. There are two main questions. One is the measurement of distances to a freely suspended mobile test object, and it has been seen, for example in the experiments of Chen *et al.* (1984) and of Spero *et al.* (1980), and elsewhere, how such measurements can be avoided by proper experimental design. The other problem is that of defining the distances which actually determine the attraction of a body. The geometrical form of the body itself must be made to the precision required in the final measurement, but that alone is not sufficient. Even if the attraction of the body is effectively equivalent to that of a sphere, the centres of volume, of mass and of attraction will not coincide unless the density is uniform. Similar problems arise with the calculation of the moment of inertia of a mechanical oscillator.

A further problem common to tests of the inverse square law and determinations of the constant of gravitation is that of extraneous forces (Chen & Cook 1990b). It arises because the forces between masses of the size that can be used in a laboratory are very small and are in fact usually very much less than the forces exerted on them by buildings, other equipment and so on. Fortunately gravitational forces add linearly so that those extraneous forces do not affect the measurements provided they are constant, but they are often not constant, and unless carefully monitored may produce significant errors. Extraneous gravitational forces will of course be more troublesome, the more precise that other aspects of an experiment can be made.

Tests of the weak principle of equivalence are inherently far more sensitive than tests of the inverse square law because they depend on detecting a small difference between large inertial and gravitational forces, and the effects of extraneous attractions are now to be compared with the total attraction of the Earth or the Sun rather than that of a small mass in the laboratory. Again, the fields of the Earth and Sun are effectively uniform, which means that the details of the form of a test mass and of its potential are of much less significance. However, at sensitivities of parts in 10^{12} those considerations cannot be ignored, as indeed the work of Roll *et al.* (1964) demonstrates.

When all has been said and done about the experimental design and metrology, the fact that gravitational forces are small compared with all others encountered in the normal laboratory is the dominant factor in experiments on gravitation. The experimenter must search carefully for effects of change of temperature (usually the most insidious), and for those of electrical and magnetic fields. They may appear as correlations between gravitational effects and temperature, time of day and other parameters, sometimes without obvious physical cause. Even when such correlations have been allowed for, there may be, indeed there most often are, internal inconsistencies without explanation. Thus, Roll *et al.* (1964) found periodicities in their results other than the 24th period that would correspond to a failure of weak equivalence, while the different values of *G* for different materials that Heyl (1930) obtained are inconsistent with all other work. These and other instances, whether of internal inconsistencies or of 'effects' that have been reported as deviations from the Newtonian law of general relativity, indicate that at the levels of mechanical sensitivity to which one must work in useful experiments on gravitation, it is difficult, if not impossible, to characterize an experiment in full and sufficient detail. The mathematical model used to calculate the results does not correspond closely enough to the actual (unknown) physical state of affairs.

An unhappy aspect of laboratory experiments in general is that for one reason or another, few of them seem to have been exploited as fully as they might have been. The final observations of Roll *et al.* (1964) were disturbed by building work nearby so that it was not possible to obtain runs long enough for a proper Fourier analysis, and similarly the observations of Braginsky & Panov (1972) were too short for Fourier analyses to be done on them. The experiments of Chen *et al.* (1984) would probably have given better results had they been performed in a more isolated site. Similar comments could be made about most of the work that has been described in this book; once the experiments have been done, it becomes clear how they might have been done better.

The scope for more sensitive laboratory experiments on gravitation, making use of the lessons of past work, is therefore plain. It is likely also that some experiments could be better done in space vehicles (Berman & Forward, 1968), although some caution is advisable. Until some possible schemes are actually tried it is not clear that they would in fact be better than Earth-bound experiments – for example, could a satisfactory torsion pendulum be devised in the absence of the control of the gravity field of the Earth, and what effect would the accelerations that control the motion of a space vehicle have on the apparent gravitational forces? In any case, it takes a long time to plan and effect any space experiment and perhaps the student of gravitation would be better to spend time and money on improvements to terrestrial experiments.

Would it be worth while? We seem to have reached a state now where all the deviations from Newtonian gravity or general relativity that were found or thought to have been found in the last two decades have turned out to be illusory. We do not need to know the constant of gravitation to better than a few parts in ten thousand. So far we have no firm predictions from grand unified theories of gravitation and everything else, and it might be argued that we should now accept that general relativity is the correct and only description of gravitation until someone produces new predictions. Yet physics is an empirical science and the present limits of experiment are a challenge to the experimenter. It has often happened that when the precision of some measurement has been improved, new physics has been revealed, often in unexpected ways. What would we find if we could check the weak principle of equivalence to one part in 10^{13} or determine the constant of gravitation to one part in a million? Perhaps nothing new, just the greater assurance that general relativity is right, but then again, something might turn up unrelated to gravitation. It does seem worth while to pursue these measurements further even if at present we may be reasonably convinced that general relativity is the theory of gravitation.

References

Abramowitz, M. & Stegun, I. A., 1964. *Handbook of Mathematical Functions* *Nat. Bur. Stds. Appl. Math. Series*, **55** (Washington, DC: US Govt. Printing Office).

Acharya, R. & Hogan, P. A., 1973. Equivalence principle of massive Brans–Dicke and Einstein theories of gravitation. *Lett. Nuovo Cim.*, **6**, 668–72.

Airy, G. B., 1855. On the computation of the effect of the attraction of mountain masses as disturbing the apparent astronomical latitude of stations in geodetic survey. *Mon. Not. Roy. Astronom. Soc.*, **16**, 42–3.

Airy, G. B., 1856. Account of pendulum experiments undertaken in the Harton Colliery for the purpose of determining the mean density of the Earth. *Philos. Trans. Roy. Soc.*, **146**, 297–342.

Anandan, J. 1984a. New relativistic gravitational effects using charged-particle interferometry. *Gen. Rel. Gravitation*, **16**, 33–41.

Anandan, J., 1984b. Effect of Newtonian gravitational potential on a superfluid Josephson interferometer. *Phys. Rev.*, B**130**, 3712–21.

Anderson, J. D., Esposito, P. B., Martin, W., Thornton, C. L. & Muhleman, D. O., 1975. Experimental test of general relativity using time-delay data from *Mariner* 6 and *Mariner* 7. *Astrophys. J.*, **200**, 221–33.

Austin, L. W. & Thwing, C. B., 1897. An experimental research on gravitational permeability. *Phys. Rev.*, **5**, 294–300.

Baily, F., 1842. An account of some experiments with the torsion rod for determining the mean density of the Earth. *Mon. Not. Roy. Astronom. Soc.*, **5**, 197–206.

Baily, F., 1843. Experiments with the torsion rod for determining the mean density of the Earth. *Mem. Roy. Astronom. Soc.*, **14**, 1–120 and i–ccxlvii.

Basset, A. B., 1886. On a method of finding the potential of circular discs by means of Bessel's functions. *Proc. Camb. Phil. Soc.*, **5**, 425–33.

Beams, J. W., Kuhlthau, A. R., Lowry, R. A. & Parker, H. M. 1965. New method for measuring the gravitational constant G. *Bull. Amer. Phys. Soc.*, **11**, 249.

Beams, J. W., 1971. Finding a better value for G. *Physics Today*, **24**, 35–6.

Berger, J. & Levine, J., 1974. The spectrum of Earth strain from 10^{-8} to 10^{2} Hz. *J. Geophys. Res.*, **79**, 1210–14.

Berman, D. & Forward, R. L., 1968. Exploitation of space for experimental research. *Amer. Astronaut. Soc., Sci. Tech. Ser.*, **24**, 95–115.

Bessel, F. W., 1832a. Versuche über die Kraft mit welcher die Erde Körper von verschiedene Beschaffenheit anzieht. *Ann. Phys. Chem. (Poggendorf)*, **25**, 401–17.

Bessel, F. W., 1832b. *Mémoires relatifs à la pendule*, volumes 4 and 5.

Bessel, F. W., 1889. *Collection de mémoires relatifs à la physique*, publiés par la Société française de Physique, **5**, 72–133.

Bondi, H., 1957. Negative mass in general relativity. *Rev. Mod. Phys.*, **29**, 423–8.

Bouguer, P., 1754. Sur la direction qu'affectent les fils-a-plomb. *Hist. Acad. Sci., Mem. Math. Phys.*, **1–10**, 150–68.

Boys, C. V., 1889. On the Cavendish experiment. *Proc. Roy. Soc.*, **46**, 253–68.

Boys, C. V., 1895. On the Newtonian constant of gravitation. *Philos. Trans. Roy. Soc.*, A**186**, 1–72.

Braginsky, V. B., 1968. Classical and quantum restrictions on the detection of weak actions on a macroscopic oscillator. *Sov. Phys.-JETP*, **26**, 831–4.

Braginsky, V. B., Caves, C. M. & Thorne, K. S., 1977. Laboratory experiments to test relativistic gravity. *Phys. Rev.*, D**15**, 2047–68.

Braginsky, V. B. & Manukin, A. B., 1977. *Measurement of Weak Forces in Physics Experiments* (Engl. transl. ed. D. H. Douglass) (Chicago and London: University of Chicago Press).

Braginsky, V. B. & Panov, V. I., 1972. Verification of the equivalence of inertial and gravitational mass. *Sov. Phys.-JETP*, **34**, 463–76.

Braginsky, V. B., Rudenko, V. N. & Rukman, G., 1962. An experimental investigation of the effect of an intermediate medium on gravitational interaction. *Sov. Phys.-JETP*, **16**, 36–41.

Braun, C., 1897a. Die Gravitationskonstante, die Masse und mittlere Dichte der Erde. *Denkschr. Kaiserl. Akad. Wiss. (Wien) Math. naturwiss. Kl.*, **64**, 187–258.

Braun, C., 1897b. A new determination of the gravitation constant and the mean density of the Earth. *Nature*, **56**, (1897), 56 and 198.

Caputo, M., 1962. Un nuovo limite superiore per il coefficiente di assorbimento della gravitazione. *Atti. Acad. Linc., Rend.*, **32**, 509–15.

Carlini, F., 1824. Osservazioni della lunghezza del pendolo semplice fatte all'altezza di mille tese sul livello del mare. *Eff. Astron. di Milano*, 28–40.

Cavendish, H., 1798. Experiments to determine the density of the Earth. *Philos. Trans. Roy. Soc.*, **88**, 469–526.

Chan, H. A., Moody, M. V. & Paik, H. J., 1982. Null test of the gravitational inverse square law. *Phys. Rev. Lett.*, **49**, 1745–8.

Chapman, S. & Bartels, J., 1940. *Geomagnetism*. 2 vols. (Oxford: The Clarendon Press).

Chen, Y. T., 1982. The gravitational field inside a long hollow cylinder of finite length. *Proc. Roy. Soc.*, A**382**, 75–82.

Chen, Y. T., 1984a. Non-linear effect in the determination of G by the swing method. *Phys. Lett.*, A**106**, 19–22.

Chen, Y. T., 1984b. Initial work of a new experiment to determine G by mechanical resonance method. In *Proc. 7th Int. Conf. Atomic Masses. Fund consts*, ed. O. Klepper (Darmstadt: Technische Hochschule), 705–11.

Chen, Y. T. & Cook, A. H., 1989. Mathematical behaviour around the stationary point of a gravitational ring. *Phys. Lett.*, A**138**, 378–80.

Chen, Y. T. & Cook, A. H., 1990a. Thermal noise limitations in torsion pendulum experiments. *J. Phys. (Class. Quantum. Grav.)*, 7, 1225–39.

Chen, Y. T. & Cook, A. H., 1990b. How weak a gravitational force can be measured in the laboratory? *Rend. Fis. Accad. Lincei*, ser. 9, **1**, 373–8.

Chen, Y. T., Cook, A. H. & Metherell, A. J. F., 1984. An experimental test of the inverse square law of gravitation at range of 0.1 m. *Proc. Roy. Soc.* A**394**, 47–68.

Cohen, E. H. & Taylor, B. N., 1986. *The 1986 Adjustment of the Fundamental Physical Constants (Codata Bull. 62)* (Oxford & New York: Pergamon).

Cook, A. H. 1987a. The inverse square law of gravitation. *Contemp. Phys.* **28**, 159–75.

Cook, A. H. 1987b. Experiments on gravitation. In *300 Years of Gravitation*, ed. S. W. Hawking & W. Israel (Cambridge: Cambridge University Press), 50–79.

Cook, A. H. 1988a. Experiments on gravitation. *Rep. Prog. Phys.*, **51**, 707–57.

Cook, A. H. 1988b. *The motion of the Moon* (Bristol and Philadelphia: Adam Hilger).

Cook, A. H. & Chen, Y. T., 1982. On the significance of the radial Newtonian gravitational force of the finite cylinder. *J. Phys. A. Math. Gen.*, **15**, 1591–7.

Cornu, A. & Baille, J. B., 1873. Détermination nouvelle de la constante de l'attraction et de la densité moyenne de la Terre. *C. R. Acad. Sci. (Paris)*, **76**, 954–5.

Cornu, A. & Baille, J. B., 1878. Sur la mesure de la densité moyenne de la Terre. *C. R. Acad. Sci. (Paris)*, **86**, 699–702.

Cranshaw, T. E. & Schiffer, J. P., 1964. Measurement of the gravitational redshift with the Mössbauer effect. *Proc. Phys. Soc.*, **84**, 245–56.

Cruz, J. Y., Harrison, J. C., Speake, C. C., Niebauer, T. M., McHugh, M. P., Keyser, P. T., Faller, J. E., Mäkinen, J. & Beruff, R. B., 1991. A test of Newton's inverse square law of gravitation using the 300-m tower at Erie, Colorado. *J. Geophys. Res.*, **96**, B12, 20 073-20 092.

Dabbs, J. W. T., Harvey, J. A., Page, D. & Horstmann, H., 1965. Gravitational acceleration of free neutrons. *Phys. Rev.*, B**139**, 756–60.

de Boer, H., 1984. Experiments relating to the gravitational constant. *Proc. 2nd Precision Meas. Conf., Gaithersburg 1981*, ed. B. N. Taylor & W. D. Phillips. *Nat. Bur. Stds. Sp. Publ.* 617 (Washington: Dept. of Commerce), 561–72.

de Boer, H., 1987. A new experiment for the determination of the Newtonian gravitational constant. *Metrologia*, **24**, 171–4.

de Boer, H., Huars, H., Michaelis, W. & Schlimme, E., 1980. Quadrantenelektrometer also Drehmomentmesser für kleine Drehmomente. *Feinwerktech. u. Messtech.*, **88**, 237–41.

Dicke, R. H., 1964. *The Theoretical Significance of General Relativity* (New York: Gordon and Breach).

Dirac, P. A. M., 1937. The cosmological constants. *Nature*, **139**, 323.

Drever, R. W. P., 1961. A search for anisotropy of inertial mass using a free precession technique. *Phil. Mag.* (8), **6**, 683–7.

Drever, R. W. P., 1983. Interferometric detectors for gravitational radiation. In *Gravitational Radiation*, ed. N. Deruelle & T. Piran (Amsterdam: North Holland), 321–8.

Duncombe, R. L., 1956. Relativity effects for the three inner planets. *Astronom. J.*, **61**, 174–5.

Dyson, F. J., 1972. The fundamental constants and their time variation. In *Aspects of Quantum Theory*, ed. A. Salam & E. P. Wigner (Cambridge: Cambridge University Press), 213–36.

Eddington, A. S., 1920. *Space, Time and Gravitation* (Cambridge: Cambridge University Press).

Eötvös, R. V., 1891. Über die Anziehung der Erde auf verschiedene Substanzen. *Beiblätter f. Phys.*, **15**, 688–9.

Eötvös, R. V., Pekar, D. & Fekete, E., 1922. Beitrag zum Gesetze der Proportionalität von Trähigkeit und Gravität. *Ann. Phys. (Leipz.)*, **68**, 11–66.

Eötvös, R. V., Pekar, D. & Fekete, E., 1935. *Roland Eötvös Gesammelte Werke*, ed. P. Selenyi. (Budapest: Akademiai Kaido), 307–72.

Estermann, I., Simpson, O. C. & Stern, O., 1947. The free fall of atoms and the measurement of the velocity distribution in a molecular beam of cesium atoms. *Phys. Rev.*, **71**, 238–49.

Facy, M. M. L. & Pontikis, C., 1970. Gravitation – détermination de la constante de gravitation par la methode de resonance. *C. R. Acad. Sci. (Paris)* Ser B, **270**, 15–18.

Facy, M. M. L. & Pontikis, C., 1971. Gravitation – détermination de la constante de gravitation par la methode de resonance. *C. R. Acad. Sci. (Paris)* Ser B, **272**, 1397–8.

Fairbank, W. M., Witteborn, F. C. & Knight, L. V., 1962. An experiment to determine the gravitational force on free electrons and positrons. *Science*, **144**, 562.

Fischbach, E., Sudarsky, D., Szafer, A., Talmadge, C. & Aronson, S. H., 1986. Reanalysis of the Eötvös experiment. *Phys. Rev. Lett.*, **56**, 3–6.

Fischbach, E. & Talmadge, C., 1992. Six years of the fifth force. *Nature (Lond.)*, **356**, 207–15.

Fischbach, E., Gillies, G. T., Krause, D. E., Schwan, Julie, G. & Talmadge, C., 1992. Non-Newtonian gravity and new weak forces: an index of measurements and theory. *Metrologia*, **29**, 215–61.

Fix, J. E., 1972. Ambient earth motion in the period range 0.1 to 2560 sec. *Bull. Seismol. Soc. Amer.*, **62**, 1753–60.

Fomalmont, E. B. & Sramek, R. A., 1977. The deflection of radio waves by the Sun. *Comm. Astrophys.*, **7**, 19–33.

Fujii, Y., 1971. Dilation and possible non-Newtonian gravity. *Nature*, **234**, 5–7.

Fujii, Y., 1972. Scale invariance and the gravity of hadrons. *Ann. Phys. (New York)*, **69**, 494–521.

Gibbons, G. W. & Whiting, B. F., 1981. Newtonian gravity measurements impose constraints on unification theories. *Nature*, **291**, 636–8.

Gillies, G. T., 1987. The Newtonian gravitational constant. Supplement to *Metrologia*, **24**, 1–56.

Goldman, T., Hughes, R. J. & Nieto, M. M., 1987. Gravitational acceleration of antiprotons and positrons. *Phys. Rev.*, **D36**, 1254–6.

Hague, B. & Foord, T. R., 1971. *Alternating Current Bridge Methods*, 6th edn (London: Pitman).

Harrison, J. C., 1963. A note on the paper 'Earth-tide observations made during the International Geophysical Year'. *J. Geophys. Res.*, **68**, 1517–18.

Haugan, M. P., 1979. Energy conservation and the principle of equivalence. *Ann. Phys. (New York)*, **118**, 156.

Haugan, M. P., 1985. Post-Newtonian arrival time analysis for a pulsar in a binary system. *Astrophys. J.*, **296**, 1–12.

Haugan, M. P. & Will, C. M., 1976. Weak interactions and Eötvös experiments. *Phys. Rev. Lett.*, **37**, 1–4.

Hawking, S. W. & Israel, W., 1987. *300 Years of Gravitation* (Cambridge: Cambridge University Press).

Heyl, P. R., 1930. A redetermination of the constant of gravitation. *J. Res. Nat. Bur. Stds.*, **5**, 1243–90.

Heyl, P. R. & Chrzanowski, P., 1942. A new redetermination of the constant of gravitation. *J. Res. Nat. Bur. Stds.*, **29**, 1–31.

Hirakawa, H., Narihara, K. & Fujimoto, M.-K., 1976. Theory of antennas for gravitational radiation. *J. Phys. Soc. Japan*, **41**, 1093–101.

Hirakawa, H., Tsubono, K. & Oide, K., 1980. Dynamical test of the law of gravitation. *Nature*, **283**, 184–5.

Hoskins, J. K., Newman, R., Spero, R. E. & Schultz, J., 1985. Experimental tests of the gravitational inverse square law for mass separations from 2 to 105 cm. *Phys. Rev.* D**32**, 3084–95.

Hughes, V. V., Robinson, H. G. & Beltran-Lopez, V. B., 1960. Upper limit for the anisotropy of inertial mass from nuclear magnetic resonance experiments. *Phys. Rev. Lett.*, **4**, 342–4.

Hulse, R. A. & Taylor, J. H., 1975. Discovery of a pulsar in a binary system. *Astrophys. J.*, **195**, L 51–3.

Joly, J., 1890. Abstract of paper on resonance method for constant of gravitation. *Nature*, **41**, 256.

Jones, R. V. & Richards, J. C. S., 1973. The design and some applications of sensitive capacitance micrometers. *J. Phys. E. Sci. Instrum.*, **6**, 589–600.

Karagioz, O. V., Izmaylov, V. P., Agafonov, N. I., Kocheryan, E. G. & Tarakanov, Yu. A., 1976. The determination of the gravitational constant with an evacuated torsion balance. *Izvest. Acad. Sci. USSR, Phys. Solid Earth*, **5**, 351–4.

Keiser, G. M. & Faller, J. E., 1979. A new approach to the Eötvös experiment. *Bull. Amer. Phys. Soc.*, **24**, 579.

Keyser, P. T., Niebauer, T. & Faller, J. E., 1986. Comment on reanalysis of the Eötvös experiment. *Phys. Rev. Lett.*, **56**, 2425.

Kim, Y. E., Klepacki, D. J. & Hinze, W. J., 1986. The sensitivity of the mass density profile on the geophysical determination of the gravitational constant. *Phys. Lett.*, B**177**, 255–9.

Koester, L., 1976. Verification of the equivalence of gravitational and inertial mass for the neutron. *Phys. Rev.*, D**14**, 907–9.

Koester, L. & Nistler, W., 1975. New determination of the neutron–proton scattering amplitude and precise measurement of the scattering amplitudes on carbon, chlorine, fluorine and bromine. *Z. Physik*, A**27**, 189–96.

Koester, L., Nistler, W. & Waschkowski, W., 1976. Measurement of the neutron–electron interaction by the scattering of neutrons by lead and bismuth. *Phys. Rev. Lett.*, **36**, 1021–3.

Kramers, H. A., 1940. Brownian motion in a field of force and the diffusion model of chemical reactions. *Physics*, **7**, 284–304.

Kreuzer, L. B., 1968. Experimental measurement of the equivalence of active and passive gravitational mass. *Phys. Rev.*, **169**, 1007–12.

Lindsay, P., Saulson, P. & Weiss, R., 1983. A study of a long-baseline gravitational-wave antenna system. (Report for NSF under grant PHY-8109581 to MIT).

Liu, H., Zhang, P. & Qin, R., 1983. Test of Newtonian inverse square law at laboratory distances. *Kexue Tongbao*, **28**, 1328–30.

Long, D. R., 1974. Why do we believe Newtonian gravitation at laboratory dimensions? *Phys. Rev.*, D**9**, 850–2.

Long, D. R., 1976. Experimental examination of the gravitational inverse square law. *Nature*, **260**, 417–18.

Long, D. R., 1986. Vacuum polarization and non-Newtonian gravitation. *Nuovo Cim.*, B **55**, 252–6.

Long, D. R. & Ogden, D., 1974. Producing an ultra high-uniformity gravitational field. *Phys. Rev.*, D **10**, 1677–80.

Luther, G. G. & Towler, W. R., 1982. Redetermination of the Newtonian gravitational constant *G*. *Phys. Rev. Lett.*, **48**, 121–3.

Mackenzie, A. S., 1895. On the attractions of crystalline and isotropic masses at small distances. *Phys. Rev.*, **2**, 321–43.

Mackenzie, A. S., 1900. *The Laws of Gravitation: Memoirs by Sir Isaac Newton, Pierre Bouguer and Henry Cavendish. Together with Abstracts of Other Important Memoirs* (New York: American Book Co.).

MacMillan, W. D., 1958. *The Theory of the Potential* (New York: Dover Publications).

McReynolds, A. W., 1951. Gravitational acceleration of neutrons. *Phys. Rev.*, **83**, 172–3(L).

Majorana, Q., 1957. Sull'ipotesi dell'assorbimento gravitazionale. *Atti. R. Accad. Lincei, Rend.*, **22**, 392–7.

Maskelyne, N., 1775. A proposal for measuring the attraction of some hill in this kingdom by astronomical observations. *Philos. Trans. Roy. Soc.*, **65**, 495–9.

Mazur, P., 1940. On the theory of Brownian motion. *Physica*, **25**, 149–62.

Metherell, A. J. F. & Speake, C. C., 1983. The dynamics of the double-pan beam balance. *Metrologia*, **19**, 109–22.

Metherell, A. J. F., Speake, C. C., Chen. Y. T. & Faller, J. E., 1984. Optimizing the shape of the attracting mass in precision measurements of *G*. *Proc. 2nd Precision Meas. Conf., Gaithersburg 1981*, ed. B. N. Taylor & W. D. Phillips. *Nat. Bur. Stds. Sp. Publ.* 617 (Washington: Dept. of Commerce), 581–5.

Milatz, J. M. W. & Vanzolingen, J. J., 1953. The Brownian motion of electrometers. *Physica*, **19**, 181–207.

Newton, I., 1687. *Philosophiae Naturalis Principia Mathematica* (London: Joseph Streater).

Ni, W. T., 1980. Measurement of gravitational forces due to oil tank and swimming pool and the inverse square law. *Physics (Taiwan)*, **2**, 2–5.

Niebauer, T. M., McHugh, M. P. & Faller, J. E., 1987. Galilean test for the fifth force. *Phys. Rev. Lett.*, **59**, 609–12.

Nordtvedt, K., 1968. Equivalence principle for massive bodies. II: Theory. *Phys. Rev.*, **169**, 1017–25.

Ogawa, Y., Tsubono, K. & Hirakawa, H., 1982. Experimental test of the law of gravitation. *Phys. Rev.*, D **26**, 729–33.

O'Hanlon, J. 1972. Intermediate-range gravity: a generally covariant model. *Phys. Rev. Lett.*, **29**, 137–8.

Ornstein, L. S., 1927. The theory of Brownian motion for systems in which there are different potentials. *Z. f. Phys.*, **41** (11–12), 848–56.

Ornstein, L. S., Burger, H. C., Taylor, J. & Clarkson, W., 1927. Brownian movement of a galvanometer coil and the influence of the temperature of the outer circuit. *Proc. Roy. Soc.*, **115**, 391–406.

Paik, H. J., 1976. Superconducting tunable-diaphragm transducer for sensitive acceleration measurement. *J. Appl. Phys.*, **47**, 1168–78.

Paik, H. J., 1979. New null experiment to test the inverse square law of gravitation. *Phys. Rev.*, D 19, 2320–4.

Panov, V. I. & Frontov, V. N., 1979. The Cavendish experiment at large distances. *Sov. Phys.-JETP*, **50**, 852–6.

Panov, V. I. & Frontov, V. N., 1980. Experimental test of the space-variability hypothesis for the gravitational constant. In *Abstracts of the Ninth Conf. on Gen. Rel. and Gravitation*, ed. E. Schmutzer (Jena: Friedrich Schiller University), 363–4.

Phillips, P. R., 1965a. New tests of the invariance of the vacuum state under the Lorentz group. *Phys. Rev.*, B 139, 491–4.

Phillips, P. R., 1965b. Magnetic shielding in a low-temperature torsion pendulum. *Rev. Sci. Instrum.*, **50**, 1018.

Phillips, P. R., 1987. Test of spatial isotropy using a cryogenic torsional pendulum. *Phys. Rev. Lett.*, **59**, 1784–7.

Pippard, A. B., 1978. *The Physics of Vibrations, vol. 1, containing Part 1, The Simple Classical Vibrator* (Cambridge: Cambridge University Press).

Potter, H. H., 1923. Some experiments on the proportionality of mass and weight. *Proc. Roy. Soc.*, **104**, 588–610.

Potter, H. H., 1927. On the proportionality of mass and weight. *Proc. Roy. Soc.*, **113**, 731.

Pound, R. V. & Snider, J. L., 1964. Effect of gravity on nuclear resonance. *Phys. Rev. Lett.*, **13**, 539–40.

Pound, R. V. & Snider, J. L., 1965. Effect of gravity on gamma radiation. *Phys. Rev.*, B 140, 788–803.

Poynting, J. H., 1891. On the determination of the mean density of the Earth and the gravitation constant by means of the common balance. *Philos. Trans. Roy. Soc.*, A 182, 565–656.

Poynting, J. H., 1894. A history of the methods of weighing the Earth. *Proc. Birm. Phil. Soc.*, **9**, 1–23.

Prigogine, I., 1962. *Non-equilibrium Statistical Mechanics* (New York: Interscience Publishers).

Reasenberg, R. D., Shapiro, I. I., MacNeil, P. E., Goldstein, R. B., Breidenthal, J. C., Brenkle, J. P., Cain, D. L., Kaufman, T. M., Komarek, T. A. & Zygielbaum, A. I., 1979. *Viking* relativity experiment: verification of signal retardation by solar gravity. *Astrophys. J.*, **234**, L219–21.

Reich, F., 1852. Neue Versuche mit der Drehwaage. *Abh. Königl. Ges. Wiss. (Leipzig) Mat-Naturwiss. Kl.*, **1**, 385–430.

Renner, J., 1935. Experimentelle Untersuchungen über die Proportionalität von Gravität und Trähigkeit. (In Hungarian with German abstract) *Hung. Acad. Sci.*, **53** (II), 542–70.

Renner, J., 1974. Determination of gravitational constant in Budapest. In *Determination of Gravity Constant and Measurement of Certain Fine Gravity Effects*, ed. Yu D. Boulanger & M. U. Sagitov. NASA Tech. Transl. F-15, 722 (Washington, DC: US Government Printing Office).

Richardson, O. W., 1910. Gravitation and the electron theory. *Phys. Rev.*, **31**, 609–61.

Richardson, O. W., 1922. Note on gravitation. *Phil. Mag.*, **43**, 138–45.

Richarz, F. & Krigar-Menzel, P., 1898. Gravitationskonstante und mittlere Dichtigkeit der Erde, bestimmt durch Wagungen. *Ann. der Phys. (Leipzig)*, **66**, 177–93.

Roll, P. G., Krotkov, R. & Dicke, R. H., 1964. The equivalence of inertial and passive gravitational mass. *Ann. Phys. (New York)*, **26**, 442–517.

Rose, R. D., Parker, H. M., Lowry, P. A., Kuhlthau, A. R. & Beams, J. W., 1969. Determination of the gravitational constant *G*. *Phys. Rev. Lett.*, **23**, 655–8.

Sagitov, M. U., 1976. Current status of determination of the gravitational constant and the mass of the Earth. *Soviet Astronm.-AJ.*, **13**, 712–18.

Sagitov, M. U., Milyukov, V. K., Monokhov, E. A., Nazarenko, V. S., Chesnokova, T. S., Dmitrieva, T. J. & Golizinskii, L. W., 1978. On the determination of the Cavendish constant of gravitation. *Soobschcheniya Gosudarstvennegogo Astronomicheskogo Instituta im P. K. Shternberga.* 202/203, 3–18.

Saxl, E. J. & Allen, M., 1971. 1970 solar eclipse as 'seen' by a torsion pendulum. *Phys. Rev.*, D **3**, 823–5.

Scherk, J., 1979. Anti-gravity: a crazy idea? *Phys. Lett.*, B **88**, 265–7.

Schiff, L. I., 1970. Gravitation-induced electric field near a metal. II. *Phys. Rev.*, B **1**, 4649–54.

Schiff, L. I. & Barnhill, M. V., 1966. Gravitation-induced electric field near a metal. *Phys. Rev.*, **151**, 1067–71.

Shapiro, I. I., 1980. Experimental tests of the general theory of relativity. In *General Relativity and Gravitation II*, ed. A. Held (New York: Plenum Press), 469–89.

Shapiro, I. I., Counselman, C. C. III & King, R. W., 1976. Verification of the principle of equivalence for massive bodies. *Phys. Rev. Lett.*, **36**, 555–8.

Shapiro, I. I., Pettengill, G. H., Ash, M. B., Ingills, R. P., Campbell, D. P. & Dyce, R. B., 1972a. Mercury's perihelion advance – determination by radar. *Phys. Rev. Lett.*, **28**, 1594–7.

Shapiro, I. I., Smith, W. B., Ash, M. B., Ingills, R. P. & Pettengill, G. H., 1972b. Gravitational constant: experimental bound on its time variation. *Phys. Rev. Lett.*, **26**, 27–30.

Sinsky, J. & Weber, J., 1967. New source for dynamical gravitational fields. *Phys. Rev. Lett.*, **18**, 795–7.

Slichter, L. B., Caputo, M. & Hager, C. C., 1965. An experiment concerning gravitational shielding. *J. Geophys. Res.*, **70**, 1541–51.

Snider, J. L., 1972. New measurement of the solar gravitational redshift. *Phys. Rev. Lett.*, **28**, 853–6.

Southerns, L., 1910. A determination of the ratio of mass to weight for a radioactive substance. *Proc. Roy. Soc.*, A **84**, 325–44.

Speake, C. C., 1987. Fundamental limits to mass comparisons by means of a beam balance. *Proc. Roy. Soc.*, A **414**, 333–58.

Speake, C. C. & Gillies, G. T., 1987. The beam balance as a detector in experimental gravitation. *Proc. Roy. Soc.*, A **414**, 315–32.

Speake, C. C. & Quinn, T. J., 1987. Beam balance test of weak equivalence principle. *Nature*, **321**, 567–8.

Speake, C. C. & Quinn, T. J., 1988. Search for a short-range, isospin-coupling component of the fifth force with the use of a beam balance. *Phys. Rev. Lett.*, **61**, 1340–3.

Spero, R. E., Hoskins, J. K., Newman, R., Pellam, J. & Schultz, J., 1980. Test of the gravitational inverse-square law at laboratory distances. *Phys. Rev. Lett.*, **44**, 1645–8.

Stacey, F. D. & Tuck, G. J., 1981. Geophysical evidence for non-Newtonian gravity. *Nature*, **292**, 230–2.

Thomson, J. J., 1909. Ratio of weight to mass for a radioactive substance. Presidential address, Brit. Assocn Adv. Sci., Winnipeg, Canada.

Thorne, K. S., Lee, D. L. & Lightmon, A. P., 1973. Foundations for a theory of gravitational theories. *Phys. Rev.*, D7, 3563–78.

Tohes, R. V. & Richards, J. C. S., 1973. The design of some capacitance micrometers. *J. Phys. E. Sci. Instrum.*, 6, 589–600.

Torge, W., 1989. *Gravimetry* (de Gruyter).

Tsuboi, C., 1983. *Gravity* (London: Allen and Unwin).

Turner, K. C. & Hill, H. A., 1964. New experimental limit on velocity-dependent interactions of clocks and distant matter. *Phys. Rev.*, B134, 252–6.

Uhlenbeck, G. E. & Ornstein, L. S., 1930. On the theory of Brownian motion. *Phys. Rev.*, 36, 823–41.

van Flandern, T. C., 1981. Is the gravitational constant changing? *Astrophys. J.*, 248, 813–16.

von Jolly, P. H., 1881. Die Anwendung der Waage auf Probleme der Gravitation. *Ann der Physik u. Chemie (Wiedener)* NF, 14, 881–6.

Wang, M. C. & Uhlenbeck, G. E., 1945. On the theory of the Brownian motion. II. *Rev. Mod. Phys.*, 17, 323–42.

Waschkowski, W. & Koester, L., 1976. Transmission measurements on liquid probes. *Z. Naturforsch. A*, 31, 115.

Whittaker, E. T., 1935. Gauss's theorem and the concept of mass in general relativity. *Proc. Roy. Soc. A*, 149, 384–95.

Will, C. M., 1981. *Theory and Experiment in Gravitation Physics* (Cambridge: Cambridge University Press).

Will, C. M., 1987. Experimental gravitation from Newton to Einstein. In *300 Years of Gravitation*, ed. S. W. Hawking & W. Israel (Cambridge: Cambridge University Press), 80–127.

Williams, E. R., Faller, J. E. & Hill, H. A., 1971. New experimental test of Coulomb's law: a laboratory upper limit on the photon rest mass. *Phys. Rev. Lett.*, 26, 721–4.

Williams, J. G., Dicke, R. H., Bender, P. L., Alley, C. O., Carter, W. C., Currie, D. G., Eckhardt, D. H., Faller, J. E., Kaula, W. M., Mulholland, J. D., Plotkin, H. H., Poultney, S. K., Shelus, P. J., Silverberg, E. C., Sinclair, W. S., Slade, M. A. & Wilkinson, D. T., 1976. New test of the equivalence principle from lunar laser ranging. *Phys. Rev. Lett.*, 36, 551–4.

Witteborn, F. C. & Fairbank, W. M., 1967. Experimental comparison of the gravitational force on freely falling electrons and metallic electrons. *Phys. Rev. Lett.*, 19, 1049–52.

Yu, H.-T., Ni, W.-T., Hu, C.-C., Liu, F.-H., Yang, C.-H. & Liu, W.-N., 1979. Experimental determination of the gravitational forces at separations around ten meters. *Phys. Rev.*, D20, 1813–15.

Zahradnicek, J. von, 1933. Resonanzmethode für die Messung der Gravitationskonstante mittels der Drehwaage. *Phys. Z.*, 34, 126–33.

Index

Printed in the United States
By Bookmasters